THE HEALTH INSURANCE DOCTOR

LONDON: HUMPHREY MILFORD
OXFORD UNIVERSITY PRESS

THE HEALTH INSURANCE DOCTOR

HIS RÔLE IN
GREAT BRITAIN · DENMARK
AND FRANCE

By
BARBARA N. ARMSTRONG

PRINCETON
PRINCETON UNIVERSITY PRESS
1939

FIRST EDITION

SET UP AND PRINTED
IN THE UNITED STATES OF AMERICA
BY PRINCETON UNIVERSITY PRESS
AT PRINCETON, NEW JERSEY

TO

IAN ALASTAIR ARMSTRONG

FOREWORD

WITH the recent social security development in the United States has come widespread interest in health insurance. No longer just the student and the social worker, but also the legislator, the administrator, and the public at large participate in health insurance discussion. Needless to say, the doctor has a special stake in such discussion. In fact, controversy regarding health insurance tends to center about the doctor's position in the scheme. On this position rests the insured worker's protection and the government's hope of workable machinery, as well as the practitioner's assurance of a satisfactory professional life. Basic information about the situation of the health insurance doctor abroad therefore seems timely.

The study of which this book is the product rests on a foundation of twenty years of research and university teaching in the social insurance field. The study itself embodied a first-hand investigation made between May and November 1936 in Great Britain, Denmark, and France. The facts were gathered with the aid and cooperation of both the government and the organized medical profession in each of the three countries. Representative non-organization physicians were also interviewed.

The British survey included London, Manchester, and Birmingham, and also representative provincial areas (Devon, Buckingham, Derbyshire, Warwickshire, Nottingham, and the long-depressed areas of Cumberland and Durham). In Denmark both Copenhagen and also Aarhus in Jutland were visited. In France work in Nancy, in Lyons, and in a typical rural area (centering in the village of Melun) supplemented extensive inquiries in Paris. In addition to the lay directors, managers, and secretaries of the bodies administering health insurance, sixty-seven doctors were interviewed. They included not only medical men engaged in health insurance practice (both general practitioners and specialists) and executives of the organized medical profession, but also the health insurance medical administrative officers and a number of distinguished doctors in the medical schools.

Britain, Denmark, and France were selected advisedly. They are democratic countries and their experience is of special interest in America. They present contrasting health insurance histories and contrasting methods of organization of health insurance medical service.

For the reader who is unfamiliar with health insurance, a brief outline of the entire health insurance scheme precedes the text proper on the *health insurance doctor* in each of the three countries discussed.

CONTENTS

PART I: GREAT BRITAIN

I

SUMMARY OF THE BRITISH HEALTH INSURANCE SCHEME

THE doctor's all-important rôle in a health insurance system is better understood if predicated upon acquaintance with the general outline of the scheme itself. The following summary will furnish this outline for readers who are not already familiar with it.

The British Health and Invalidity Insurance system went into operation in 1912.[1] The compulsory provisions of the original act[2] covered manual wage earners without regard to earnings and non-manual employees whose salaries did not exceed £160 ($800) per annum. British wage levels were such that this covered most of the white-collar workers.

A few groups of wage earners who were not dependent upon insurable work or whose employment provided benefits equal to those of the insurance scheme were entitled to exemption from the obligations of the act. Certain government workers were excepted.

In restricting compulsory insurance to employees and excluding self-employed workers, the British act followed the usual social insurance pattern which had been developed by Continental Europe. This restriction and exclusion rested on a purely administrative base. Obviously the cobbler, small shopkeeper, or bootblack needs social insurance protection just as much as the wage earner. The wage earner, however, can be reached through his employer, who is relatively accessible to the government, whereas collection from the self-

[1] Variations were made relative to medical benefits in Ireland and administrative arrangements in Scotland. The scheme here described is that set up in England and Wales.

[2] (Stat. of 1911) 1 and 2 Geo. V, ch. 53.

employed worker involves all the administrative problems of the poll tax. The British act permitted (through voluntary insurance), but did not at this time try to compel, any of such workers to insure.

Employers, as well as their insured employees, were required to pay contributions for the support of the scheme. The Exchequer also met a part of the cost. The three contributing sources divided the burden according to the following ratios:

Males: Insured 4, Employer 3, Exchequer 2
Females: Insured 3, Employer 3, Exchequer 2

A sliding scale increased the employer's and decreased the worker's share in the lower wage groups. Employers paying less than a specified minimum were required to pay the entire workers' share as well as their own, and employers of "exempt" workers were required to contribute at the usual employer's rate. The latter provision was included to prevent employers from deriving any advantage from the hiring of "exempt" workers.

The employer was made a collecting agency for the Government. He paid both his own and his employees' share of the premium by purchasing insurance stamps which were placed in an Insurance Book for each worker. He reimbursed himself by holding back from the pay envelope the amount due him for each worker's share.

The funds brought into the national treasury by the sale of the insurance stamps each year, supplemented by the Exchequer grants, were used to finance the insurance benefits. These were of two types, *cash* and *medical*, and they were administered in different ways.

The cash benefits, in response to demands made by the great network of fraternal organizations ("Friendly Societies") which had been providing medical service and small cash benefits during illness, were obtainable by joining an Approved Society exclusively devoted to the administration of insurance cash benefits. To be "approved" a society had to be non-profit, self-governing, guarantee the standard benefits payable under the insurance act, and not reject applicants on account of age. If it had fewer than five thousand members, it was required to form a "central pooling fund" with other societies for their mutual protection. Friendly Societies were permitted to organize compulsory insurance sections into "approved" organizations. The costs of both the benefits paid out and of their administration by the Approved Society were financed by a yearly allotment from the Insurance Funds for each insured member.

Persons who either did not or could not join an Approved Society became "deposit contributors." The amount due for each of the contributors paid by the worker, his employer, and the Exchequer, became an individual credit at the post office on which, after an annual deduction to pay for the medical benefit, the deposit contributor could draw for sickness and invalidity benefit until the account was exhausted.

One of the results of the Approved Society system was that many pseudo-fraternal orders were organized as annexes and "feeders" to the Prudential, the Metropolitan, and other big industrial life insurance companies (i.e. purveyors of "funeral insurance" policies). In the course of time, these organizations became the carriers of a larger proportion of the insured workers than were served by bona fide fraternal societies.

In deference to the violent opposition of the organized medical profession to any contractual relationship with the Friendly Societies, medical benefit[3] was separately administered by specially created local committees (the "Insurance Committees") representative of the workers, the doctors, and the "public." An Insurance Committee was set up by the National Health Insurance Commissioners for each insurance area, i.e. each county and county borough. The committee received an annual allotment from the Insurance

[3] Medical benefit is given in case of industrial injury as well as on non-industrial injury and sickness. The Workmen's Compensation Act provides only cash benefits. See p. 11 *infra*.

Funds for each insured person resident in the area, and was obligated to provide and pay for the insurance scheme's medical benefit. The doctor was obligated to give the insured worker who was too ill to work a certificate of incapacity, which the worker used in making his claim for cash benefits from his Approved Society. The society, however, had no direct relations with the doctor nor any control over him. His actual contacts were with the Insurance Committee.

The British act was the first health insurance measure to throw insurance practice open to the whole medical profession and to allow free choice of doctor by the insured patient.

The cash benefits included (1) "sickness benefit" granted, after a three-day waiting period, up to twenty-six weeks of continuous incapacity;[4] (2) "disablement benefit" (half the amount of sickness benefit), an invalidity pension payable without time limit for incapacity that continued beyond the maximum duration of sickness benefit; (3) "maternity benefit," a lump sum payable on the confinement of either an insured woman or the wife of an insured man. Both contributions and cash benefits were set at flat figures that did not vary, as did the continental health insurance schemes, with the wages earned by the worker. Variations were made according to sex, and lower rates were set for "young people" considered as still in the learning period. "Qualifying periods of insurance" were set for each type of benefit except medical benefit. Medical benefit was available only to the insured man or woman, not to the dependent family.

A certain amount of "arrears" resulting from unemployment (an average of four weeks a year) was considered normal and caused no reduction in benefit. Arrears in excess of this amount caused a gradual loss of benefit, taking the form of successive reductions and final cessation of cash benefit, and, last of all, deprivation of medical benefit.

[4] Incapacity, whether or not from the same disease, is deemed a "continuation" of previous incapacity unless at least a year has elapsed since the previous incapacity.

There has been little real change in the health insurance scheme during the twenty-five years of its existence. The income figure set for determining the eligible white-collar employees was raised in 1919 from £160 ($800) to £250 ($1,250), in order to keep pace with the rise in wages and prices. It has stayed at that figure.

In 1935 there were approximately 15,500,000 insured persons and 734 centralized societies administering health insurance in England. In addition, there were twenty-five Affiliated Orders with just over five thousand branches.

Since 1918 the right to participate voluntarily in the insurance scheme has been restricted generally to persons[5] who were formerly compulsorily insured and, through change of employment or for other reason, have moved out of this insured group. Compulsory insurance was extended in 1929 to several groups of self-employed manual workers who were technically "independent contractors" rather than employees. Persons for whom they performed services were made to serve as employers for purposes of the act.

Since 1929 there has been a governmentally organized insurance carrier for deposit contributors who have been unable, because of bad health, to secure membership in an Approved Society. This "deposit contributors insurance section" functions like an Approved Society.

On the institution of old-age insurance in 1924, sickness and disablement benefits were made terminable at the age of sixty-five when the old-age pension became payable. Pensioners remain eligible for medical benefit all their lives.

Sanatorium benefit, a special service for the tubercular originally part of the regular medical benefit, has not been administered as a health insurance benefit since 1920. Instead, it has been assimilated in a comprehensive program for the treatment of tuberculosis administered by the County Councils and supported by both national and local taxes.

[5] There are a few exceptions. See National Health Insurance Act, 1936, 26 Geo. V and 1 Edw. VIII, ch. 32, sec. 3 (2), (3), (4).

In July 1919 the administration of health insurance was made one of the duties of the newly created Ministry of Health. The special government agency (the National Health Insurance Commissioners) set up by the original enactment was dispensed with. A "Medical Referee" supervisory service was set up in 1920 and has been perfected in the interim.

The contribution and benefit amounts have been revised from time to time. The shares of the employer and worker are identical in the case of male labor. The share of the employer is a little more than the worker's in the case of female labor. The ratio of the government grant to the total normal income has been substantially reduced. The required annual government grant, which before 1925 amounted to two-ninths of the cost of the insurance of male workers and one-fourth of the cost of the insurance of females, is now only one-seventh and one-fifth of the respective amounts.

The qualifying periods have been revised and are now as follows: Contributions during a half-year for sickness benefit at a reduced rate; two years for full sickness benefit and for disablement benefit; contributions during ten and a half months for maternity benefit. Medical benefit is available, as it always has been, immediately on insurance.

Arrears provisions were liberalized repeatedly. And almost continuously from 1920 on, special Exchequer grants were made to protect the health insurance rights of workers who, through prolonged unemployment caused by the depression following the World War, fell into penalized arrears. Special legislation kept reduced cash benefits available during most of this period, even in the "depressed" areas where widespread unemployment had endured for many years. Medical benefit rights have been preserved practically without interruption. The excusing of arrears resulting from bona fide unemployment now has been made a permanent part of the health insurance scheme, thus assuring the continuity of protection during all phases of future business cycles.

This maintenance of health insurance benefits has been the most conspicuous post-war health insurance service. It has meant health and economic protection to the unemployed worker, an important contribution to the public health in England. It also has saved the medical profession from the serious loss of income that bad times traditionally have brought the professional man. Government checks have gone regularly to the British health insurance doctors year in and year out. The post-war industrial adjustments which Britain had to face caused not only general depression for an unusually long period, but also a complete cessation of economic activity in certain of the mining regions. The latter condition has persisted even in the face of general economic revival. In consequence, the protection of medical benefit has been of special importance to the English physicians, as well as to the insured workers.

The cash benefits (sickness benefit and disablement benefit) still are conspicuously low as compared with unemployment insurance benefit.[6] This low level of cash benefit has been tolerated largely because supplementary amounts are commonly provided by the societies as "additional benefit," a good talking point in membership drives. More than 70 per cent of the insured workers belong to societies that provide supplementary cash benefits. Almost as many are also entitled to substantial additions to the ordinary services of medical benefit.

[6] Thus:

	Males	*Females*
Weekly Sickness Benefit	15s. ($3.75)	12s. (10s. if married)
Weekly Disablement Benefit (with no family allowances)	7s.6d. ($1.87)	6s. (5s. if married)

Unemployment insurance benefit is:

Males 17s. plus family allowance of 3s. per dependent child and 9s. for dependent wife or other person in charge of the children.

Females 15s.

(Agricultural unemployment insurance benefit, however, is only 14s. for males and 12s. for females, plus smaller family benefits.)

II

HEALTH INSURANCE MEDICAL BENEFIT

MEDICAL benefit under the British health insurance scheme affords the insured worker treatment by a "family doctor" or general practitioner and such drugs or medicaments as the doctor may prescribe. These benefits are completely financed by the insurance funds. No contribution toward the cost of the drugs is exacted from the patient.[1] No extra payment of the doctor by the worker for insured treatment is allowed, although if the doctor performs additional non-insured medical service, this must be paid for as the doctor and patient may arrange.

Medical Treatment

Just what is included in the insured medical service?[2] The rule laid down in the Medical Benefit Regulations requires, with one exception, that the practitioner give his insured patients "all proper and necessary medical services other than those involving the application of special skill and experience of a degree or kind which general practitioners as a class cannot reasonably be expected to possess . . . regard being had to what services are usually undertaken by practitioners in the locality."

The exception is related to confinement cases. Attendance during or within ten days after labor is not part of the in-

[1] The patient must furnish a bottle or container or else pay a small sum as deposit for the container. To this end, therefore, money is often passed over the counter by the insured worker at the druggist's. In "easy" times, moreover, the less thrifty workers frequently make small purchases of drugs or such goods as adhesive tape, medicine droppers, etc., rather than bother to go to the doctor for a prescription.

[2] See pp. 30*ff*. for discussion of "additional benefit" schemes.

sured service.[3] Instead, a special cash maternity benefit is provided, supposedly enabling the patient to pay either a doctor or a midwife as she desires. Antenatal examinations, however, and medical supervision of the insured woman during her pregnancy periods are part of the health insurance doctor's obligations.

The doctor is obligated to treat industrial injuries as well as non-industrial accidents and illness.[4]

The service obligations of the insurance doctor are stated, as has been said, in most general terms. Whether or not a particular operation or service falls within the insured service has caused frequent disputes. The burden of proof rests wholly with the doctor who contends that a particular disputed service is beyond his insurance obligations. If he has already performed the service, moreover, and the dispute arises over the question of extra payment, the doctor has very specific obligations of proof. The rule provides that performed service shall be deemed within the insurance obligations, unless the physician can show either special experience or special reputation among fellow medical men in this field.[5] The administering of anesthetics or other assistance at an operation on one of the doctor's patients, even when the operation is itself deemed a specialist service, is an insurance obligation, unless only a specialist could properly so assist.

Preventive treatment, such as vaccination for smallpox, inoculation against diphtheria, typhoid, or other disease, is

[3] Benefit is paid on confinement of the wife of an insured man, and double benefit on confinement of the insured woman. See p. 6 *supra*. The distinction between a confinement case and abortion is drawn as follows: If there have been twenty-eight weeks of pregnancy, it is classed as confinement, whether or not the child is born alive. If the pregnancy has lasted less than twenty-eight weeks, it is classed as an abortion unless a living child is delivered, in which case, of course, it is a confinement. Attendance at abortion is, of course, required as part of the practitioner's insurance duties.

[4] The doctor gets a special fee from non-insurance funds for emergency treatment of insured persons (not his own patients) who are hurt in traffic accidents. See Chap. III, n. 16.

[5] He must establish either that he has had extra training at a hospital or with some other appointed institution plus recent experience in the contested work or else that the doctors in his area regard him as having developed a special proficiency beyond that of the usual practitioner.

in no way distinguished from curative treatment. Whatever the doctor would do for a private patient, he is obligated to do for his insured patients.

The practitioner's duty to provide needed medical treatment is not limited in any way by the fact that the patient has brought on the condition by his own negligence or misconduct. Refusal by the patient to submit to surgery or to undergo inoculation which his doctor deems necessary does not suspend medical benefit or bring other penalty to the insured, except in the case of refusal, deemed unreasonable by the society, of very minor surgical operations ordered by the practitioner.[6]

Post-mortem examination of an insured person is not required and, if asked for, may be charged for on a private fee basis.[7] This does not mean, however, that a fee may be charged because the patient dies before the arrival of the doctor.

It is of course impossible to state exactly in administrative regulations the amount of visitation expected of an insurance practitioner in the case of chronic, long-enduring disabilities for which little or nothing ordinarily can be done. It is made clear, however, that fairly frequent supervisory visitation is required in such cases to check for possible complications that may be at least *relieved* by the physician.

Under the general rule defining the doctor's insurance obligations, local custom is important. More is expected of a country doctor, for the simple reason that he has learned to do more, because of his remoteness from hospitals and specialists. And what most general practitioners in the area can do is what is included in the insurance service obligations.

Whether extraction of teeth is a rural medical practi-

[6] Whether refusal is unreasonable is decided on appeal by the Minister of Health if the insured challenges his society's decision. cf. Danish rules in these respects, pp. 116*ff*.

[7] A death certificate must be issued, however, as part of the insurance duties. See p. 49.

tioner's obligation is a moot question in England and Wales.[8] The central administrative authorities maintain that there is nothing in the act or regulations issued thereunder that excludes such service. Accordingly, they maintain that a rural doctor who extracts teeth for private patients could not with propriety charge an insured patient for such service. Further, they contend that he should perform it at least when no dental service is reasonably available. The insurance practitioners insist that they do not accept this view, but they have not formally contested it. The fact that a special "dental benefit" is largely available as an *additional benefit*[9] (voluntarily granted by the societies) probably accounts for the fact that the question has not been actually litigated.

It is conceded by the medical group that such disability as hemorrhage of the gums is clearly entitled to treatment by the health insurance doctor when dental assistance is not available in the environs.

Disputes over the limits of general practitioner service are referred in the first instance to a county administrative committee of doctors, the Medical Committee, chosen by vote of all the doctors in the area, including non-insurance as well as insurance doctors. This committee reports to the County Insurance Committee, which includes doctors, administrative officials, and representatives of the insured workers. Should the Insurance Committee disagree with the Medical Committee, the matter goes for final decision to three referees appointed by the Minister of Health.[10] Two of these referees must be doctors, the third a lawyer in active practice. Thus the limits of the insured service have been defined largely by medical men.

[8] In Scotland it has been formally decided, on contest, by referees that extraction of teeth is outside the scope of medical benefits. See Harris and Sack, *Medical Insurance Practice* (4th ed., 1937), British Medical Association, London, p. 62.

[9] See pp. 30*ff*.

[10] The Minister may send the case to the referees even when the two local committees agree.

The right to free drugs, which is a part of the medical benefit, is placed for administration primarily in the hands of the attending doctor. The cost of the drug which he deems necessary to the treatment of the condition of his patient is irrelevant: the patient's need, and not the cost of the material, determines the range of the drug benefit.[11] The doctor must not, however, prescribe an expensive medicine if a less costly and equally efficacious one is available. He may not cater unduly to the palates of his patients by ordering excessive amounts of "tasty" flavoring matter.[12]

Drugs that are immediately necessary and cannot wait the filling of a prescription must be dispensed by the doctor. He *may* also dispense the drugs which ordinarily are administered by him personally. He is reimbursed for these (and compensated for the dispensing) by the insurance funds. Country doctors more than a mile from a drug store may furnish all drugs, instead of merely prescribing them; they must do so if the Insurance Committee finds that the insured patients have no other reasonable means of obtaining them.

Occasionally disputes arise over borderline commodities, which may be deemed food, beverages, disinfectants, or toilet preparations, instead of drugs.[13] For instance, certain prescriptions for concentrated milk products used for their protein content in anemia cases were challenged by a druggist on the ground that they were food rather than medicine. The Insurance Committee, after consultation with a Medical

[11] The drugs are, of course, limited to medicines required in connection with medical treatment which is provided under the insurance act.

Patent medicines are not absolutely interdicted but the burden of proof is on the doctor who prescribes them to show that they have therapeutic value which cannot be obtained by non-proprietary remedies. That the patient prefers the proprietary article is not accepted as such proof.

[12] See pp. 63*ff.* for further discussion and for description of procedure in cases of alleged over-prescribing, i.e. prescribing unnecessary or unnecessarily expensive drugs.

[13] The Advisory Medical Committee published a report in 1929 (with later addendum) for the guidance of practitioners. This lists substances in three classifications: 1. Never a Drug. 2. Always a Drug. 3. Sometimes a Drug. General rules are appended to aid the doctor. See Appendix 17 of Harris and Sack, *Medical Insurance Practice*, for useful extracts from this report.

Committee representing insurance doctors in the area, agreed with the druggist. The dispute ultimately was resolved, as are all such disputes, by appeal to the Minister of Health, who in this instance held with the prescribing doctor that the compound was within the drug benefit.

In addition to medicine proper, certain chemical reagents (such as Fehling or Benedict Solutions for regulating diabetic treatment) are included, as are the appliances listed in the Official Regulations. The list includes the things ordinarily needed in carrying on treatment supervised by a general practitioner, such as bandages of all varieties, adhesive and other types of plaster, syringes, hypodermic needles, droppers, lints, ice bags, surgical jackets, splints, and similar commodities.[14]

Included in the obligations of the insurance doctor is the furnishing of certificates of incapacity and capacity. These certificates guide the organizations that furnish the cash benefit to which health insurance entitles the worker when illness disables him for work. It is stressed in Britain that this certification duty is part of the physician's treatment obligations and not an addition to them. The British Medical Association, as well as the medical administrators of the insurance scheme, point out that a sick man does not get well by drugs alone. They stress the fact that food and shelter for him and the peace of mind that comes from knowledge that his family have their maintenance secured are important elements in his recovery. The cash benefit, they maintain, is a vital part of treatment itself. In turn, the attending doctor's certification, which enables the worker to get his cash benefit, they deem a normal obligation of the doctor, comparable to his duty of prescribing the free drugs which are part of medical benefit. "Certification is a part of treatment" is a cornerstone in the British health insurance scheme.

[14] See *Great Britain Statutory Rules and Orders*, 1936, no. 1415, 2nd schedule, for current list of appliances, etc., supplied as part of medical benefit.

In addition to treatment obligations, the insurance doctor must advise and direct a patient who needs specialist services. If the services needed are provided by the public authorities, he must take all reasonable active steps to enable his patient to obtain them.

III

THE HEALTH INSURANCE DOCTOR

TO be a health insurance doctor in Britain a practitioner must be willing to work on the terms offered under the health insurance act. In addition, he must, like any doctor in "private practice," attract patients who wish to have him for their doctor.

Any man or woman "on the medical register," i.e. admitted to practice, may put his name on the county "panel." This is the list of medical men who offer their services as health insurance doctors. If a doctor practises in more than one insurance area, he of course joins more than one panel. The insured worker may choose any doctor on the local panel and, if the patient is acceptable to the doctor, the worker's name is put on the doctor's medical list. Obviously, these medical lists are not lists of actual patients in the sense of sick people needing medical attention. They represent merely potential patients who look to the doctor in question for medical service if and when they need it.[1]

The worker receives a medical card (with an identification cipher) from his Insurance Committee, i.e. the local administrative body. He chooses a doctor by presenting his medical card to a practitioner whose name is on the panel. If the doctor accepts him, he signs the medical card and sends it to the Insurance Committee. The committee puts the worker's name on the doctor's list and returns the medical card to the worker.[2] If the doctor wishes to reject a worker

[1] Staff physicians of hospitals, asylums, etc., are permitted by the Insurance Committees to serve as insurance doctors for the employees of the institution only. The lists of such doctors are called "limited lists."

[2] A medical index register of workers eligible to medical benefit is kept by each Insurance Committee and the Committees clear through a Central Index Register, which keeps the Committees informed of the state of each worker's contribution record.

who has chosen him, he is free to do so. He must, however, give him any treatment of which he stands in immediate need. In addition, he must direct him to other doctors in the neighborhood to whom he may apply and also notify the committee that he has rejected the applicant.

The insured person who wishes to change his doctor may, with the consent of the practitioners involved, do so at any time. Even without such consent, the worker may change four times a year—at the beginning of each quarter. He must, however, give a month's notice of his intention to make such change, i.e. a change without acquiescence of the doctor from whose list he is removing himself.

The doctor who has accepted an insured worker has in turn the right to make a subsequent change. With the consent of the patient and another doctor, he may remove the worker's name from his list at any time and have him transferred. Even without the consent of his patient, he may do so on fourteen days' notice. The name is stricken off before that time if accepted on the list of another doctor.

The experience of representative areas may serve to illustrate the extent of "changes" of doctor under the health insurance scheme. A negligible number are transferred at the request of the practitioner. Even in the areas with several hundred thousand insured, only ten to twenty per year are removed from one list to another by notice from the doctor.

In quoting material, it must be borne in mind that the statistics kept by most insurance committees of "changes of doctor at the patient's request" are not properly comparable. Some counties segregate changes made because of removal from the doctor's neighborhood, while others lump together such changes with those made because of dissatisfaction with the attending doctor. Some counties do not distinguish between changes made "by consent" and those made "on notice."

Thus with an insured population of about 330,000, the changes in the Manchester area from 1932 to 1936 were as follows:

Year	*Number of Changes*[3]
1932	12,572
1933	13,680
1934	13,566
1935	13,109

The Insurance Committee estimates, however, that 80 per cent of these "changes" of doctor were caused by the patient's removal[4] to a residence more than two miles from his former doctor's office, and only 20 per cent (2,620 cases) involved transfer solely because of dissatisfaction with the attending practitioner. This constitutes well under 1 per cent of the insured persons in the area. Figures were not kept to indicate how many of these changes were by "consent" and how many by "notice."

In Derbyshire the changes recorded ran above the Manchester figure, averaging 4,000 to 5,000 a year for an insured group of 250,000—i.e. from 1.60 per cent to 2 per cent of the group. Two-thirds of these transfers were by "notice" and one-third by "consent." How many were caused by change of residence rather than a real desire for a change of doctor is not recorded, and the committee makes no estimate of this figure.

In London, with approximately 1,900,000 insured persons, the transfers "otherwise than by consent" were far more numerous. They averaged approximately 70,300 a year in the years 1932-1935 inclusive, which is more than 3.5 per cent of the insured group.

In Buckinghamshire annual changes averaged 965 in the years 1933-1935 inclusive, with an average of 105,000 insured. This is less than 1 per cent of those eligible to medical benefit. The changes, moreover, include "consent" as well as

[3] Figures furnished by the Insurance Committee of Manchester.

[4] In such a situation the insured could, under a rule of the local Insurance Committee, transfer to another doctor without notice or consent of his former practitioner.

"notice" transfers and transfers because of change of residence.

In the twenty-five years during which the British health insurance scheme has been in operation, the rules governing the right to change doctors and the right of the doctor to remove patients from his list have been frequently altered.[5] From the beginning, change with the consent of all persons involved (the two doctors and the insured person) has been permitted *at any time*. Change against the will of the attending physician was at first permitted only once a year. This period was reduced several times and then abolished altogether, changes being permitted at will. A two weeks' notice rule was later adopted, and finally, in 1931, the provisions allowing a non-consent change to be made only four times a year replaced the fourteen days' notice rule. The present rule was the result of a conviction current in 1931 on the part of the administrative authorities that it would check hasty and "tempery" changes and also would relieve the doctor of possible pressure from some of his patients to be generous in "certifying" them for their cash benefit.[6]

The organized medical profession has remained insistent that change at any time should be permitted, so that there will be no difference whatsoever in the relations of doctor and patient under the insurance scheme and in private practice.

On the whole, however, the Insurance Committee secretaries seem to attach little importance to the "notice" requirement in changing doctors. It is their belief that approximately the same number of persons move about from doctor to doctor whether or not notice is required and regardless of the notice period. They believe that the notice requirement is not justified on the score of preventing "won't works" from annoying doctors for undeserved certificates of incapacity. Most of them are emphatic that the extent of deliberate malingering by patients is small if not

[5] For discussion of arrangements in Manchester and Salford see the next chapter.

[6] See pp. 58*ff.* for complete discussion of certification.

negligible. Several suggested that the only material result of the quarterly notice rule was a "bunching" of the changes that otherwise would be spread out evenly over the year. One committee (for Buckingham) had rather interesting evidence that the greatest percentage of transfers had not coincided with the period when patients could change doctor at will!

CHANGES EFFECTED IN SECOND AND THIRD QUARTER OF THE YEAR[7]

	At Will	*On 14 Days' Notice*	*Every Quarter on Month's Notice*
	1927	1930	1933
2nd Quarter	271	353	230
3rd Quarter	196	339	190

The secretary of the Borough of Derby Insurance Committee reported as follows: "When the change could operate at any time, the number of changes was undoubtedly less than it is now, but so far as this area is concerned, the changes do not vary a great deal, the main reason being that we are not overstocked with doctors and choice is consequently limited." With an insured population of 105,000, the changes each year since 1930 have varied between 506 and 555 per annum.[8]

The medical benefit regulations provide that, except in unusual circumstances and with special permission from the authorities, a doctor practising alone may not have more than 2,500 persons on his list. If two or more health insurance doctors practise in partnership, the list of one of the partners may reach (but not exceed) 3,000, so long as the average on the lists of the partners is not over 2,500. If a practitioner or a partnership employs one or more permanent assistants, an additional 1,500 names per assistant are permitted. As has been previously remarked, these lists total *potential*, not actual patients.

[7] Information furnished in June 1936 by Mr. Woods, secretary of the Insurance Committee for Buckinghamshire.

[8] Information furnished in June 1936 by Mr. Irving, secretary of the Insurance Committee of Derby Borough.

A doctor's decision to place his name on the health insurance panel is, of course, not irrevocable. If he wishes thereafter to leave insurance practice, ordinarily he may have his name withdrawn from the panel on proper notice.[9] The period of notice is at most three months and may be any shorter period agreed to by the Insurance Committee. The practitioner who wishes to withdraw because the terms of insurance service are altered, either as to duties or remuneration, may be held for two months at most.[10] The notice requirements are obviously intended to give time for arranging substitute medical service for the insured workers involved.

Withdrawals from insurance work apparently are not very frequent. They result most frequently from the doctor's removal from the area, and as a rule every effort is made by the authorities to accommodate the doctor who presses for immediate release. Some withdrawals, however, are made as a result of "sale" of practice or good will. This has always been customary in Britain both within and without the insurance system. These sales are facilitated by the procedure described in the next paragraph. There is occasional complaint that a few doctors have taken advantage of this custom of selling practices by making a business of working up an insurance practice just to sell it.

On the death of an insurance doctor, the committee may permit a representative of the estate to appoint a temporary "deputy" physician, on salary paid by the estate. He may handle the insurance practice of the deceased for not more than two months, during which period the practice is sold. Names are transferred to another doctor's list during this

[9] If, however, he has been accused of misconduct in connection with his insurance practice, he may retire to private practice only with permission of the Minister of Health and on such conditions as the Minister may impose. This, of course, is to prevent a doctor who is guilty of misconduct from saving himself from possible loss of privilege of health insurance practice by voluntary "withdrawal" when under fire. See pp. 62*ff.* for discussion of disciplinary control of the doctor.

[10] Provided he acts within a month of notice of the change in terms.

two months' period only on request of the patient. Thereafter, on request of the representative, the committee must notify each insured worker on the list (as on the request of a withdrawing practitioner who "sells" his practice) that a certain named practitioner—the doctor who has "bought" the practice—is willing to accept him for treatment, and that, in the absence of formal request within a month for other arrangements, he will be deemed to be on this practitioner's list. Notification for other arrangements involves well under 5 per cent of the practices that are "sold."

Additions and Removals by the Insurance Committee

Names may be added to medical lists by the local insurance committees in two sets of circumstances: (1) when the worker fails to choose a doctor, and (2) when the doctor chooses to reject a patient.

Not every insured person exercises his privilege of choosing a doctor. About 2 to 3 per cent on the average (and in large cities more than 5 per cent)[11] of the workers in an insurance area are not on any doctor's list. Usually this happens either because they never select a doctor or because the doctor they have selected withdraws from service without naming a successor. Most counties merely divide among all the panel doctors in proportion to their lists the money available for the persons who are not on any doctor's list.[12] In some of the counties, however, after three months' time, the Insurance Committees assign such men and women to doctors who signify their willingness to receive allocations. Men building up a practice often welcome this addition to their responsibilities and income.[13]

[11] Thus out of 255,415 on the insured register in Derbyshire (county area), 250,121 were on doctors' lists by their own action, 5,294 were not on doctors' lists (April 1936). Of 1,887,517 on insured register in London, 100,000 not on doctors' lists by their own action (March 1936).

[12] See pp. 78*ff*.

[13] The London Insurance Committee commenced only in October 1934 to "allocate." They regard the present system as still tentative and experimental.

The relative unpopularity of "allocation" is explained as follows. It is very difficult to keep the lists of insured persons accurate and up to date, especially in the cities where there is a substantial group of floaters who come in and out of the area without notifying the committee. This, of course, complicates the allocation problem, for a doctor may be assigned names of persons who no longer live in the district. The doctor thus gets the right to payment because of responsibility for the care of non-existent patients. Most of the committees, therefore, favor the system of merely dividing among all the doctors the funds available for the number listed as eligible but on no doctor's list, rather than allocation of individual workers to the medical lists of individual doctors.[14]

Years of experience have convinced the secretaries of the Insurance Committees that at any given time the number of names of insured persons who in reality have left the area (without notifying the insurance authorities) is about balanced by the number who have come in and have not had themselves added to the list of workers eligible for medical benefit.

Occasionally workers who may be undesirable patients fail to find any doctor who will accept them. Yet the law entitles all insured workers to medical service. If the worker who is rejected by the practitioner or practitioners of his choice asks the Insurance Committee to find him a doctor, the committee may assign him arbitrarily. Even a doctor who has refused him may have to accept him on assignment by the committee, although such an arrangement would be made only in rural areas where all available doctors have refused

[14] According to the report of the Chief Medical Officer of the Ministry of Health for 1936 there were 420,154 persons in England and Wales who had not selected their doctor or made other arrangements for receiving medical treatment. This is approximately 2.6 per cent of those (16,320,000) he lists as eligible to medical benefit. His figures suggest, however, that the number he refers to as "having selected" includes those allocated by the Insurance Committees as well as those who have actually chosen of their own initiative. (See *Annual Report* of the Chief Medical Officer of the Ministry of Health for the year 1936, p. 75.)

an applicant. The "nuisances" are then parcelled out by rotation so as to distribute them as fairly as possible among the doctors concerned. The medical men feel no injury from this situation, as they accept as inevitable and proper their collective responsibility for treatment of the insured population.

Refusal by the doctor to accept a patient is very infrequent. For example, the Manchester Committee records only about a dozen rejections a year for 320,000 insured. In Birmingham, with 475,000 insured, the committee reports from twelve to twenty such refusals a year.

The Insurance Committee may remove insured persons from a doctor's list if, after a warning that he has exceeded the maximum number allowed, the doctor does not adjust the matter himself. The committee also removes from the doctor's list the names of persons who, with the intention of staying three months or more, move from the area in which the doctor practises. Workers who go away for less than three months remain on their home doctor's list and obtain any needed medical treatment during their stay elsewhere on a "temporary residence"[15] basis or, if it is merely an overnight stay, according to the rules in some counties, on an "emergency" basis.

The latter arrangement is intended primarily for cases of accident or other sudden emergency. Treatment may then be demanded by an insured person of any insurance doctor if the worker's own physician is not available. There is no fixed definition of a "sudden emergency" or of the phrase "not available"; both are interpreted with practical com-

[15] When a worker applies to a doctor on moving into a new insurance area, the doctor signs the "B" part of his medical card in case of a "temporary resident" and the "C" part if it is a permanent removal, i.e. for more than three months. The card is sent to the new doctor's committee and cleared through the Central Index Register, which keeps the committee informed of the movements of the workers on their medical register. If the worker claims to be insured and does not present his medical card, the doctor may charge a fee. The worker, if really entitled to medical benefits, can recover this from the Insurance Committee, which in turn holds the amount out of the doctor's next check.

mon sense. If a sudden urgent situation such as a hemorrhage occurs and there is no reasonable probability that the worker's own doctor can get to the spot without prejudicial delay, any insurance doctor must respond when called upon for help. Such service is specially paid for out of the Insurance Funds.[16]

The doctor has an obligation toward "temporary residents" identical with that which he owes a regular resident who selects him for his doctor. The practitioner to whom he applies must either accept him for the period of his stay or else give needed temporary treatment pending his acceptance by another doctor. Treatment of temporary residents, like emergency treatment, is paid for on a special basis.[17]

The doctor is permitted to arrange with another practitioner to act as his deputy in performing certain services which, while within the insurance contract, are services which he is unwilling to perform. Thus, for example, he may and sometimes does prefer to delegate certain surgical work to another man. Every doctor is required, moreover, to provide for the handling of his practice during his temporary absence or when, because of conflicting professional duties, he cannot meet all his insurance obligations. He must, of course, inform his patients on these points. He must also notify the Insurance Committee if he has any standing arrangements. Failing such, he must not leave his practice for more than a week without advising the committee of his proposed absence and furnishing the name of the practitioner who will be responsible as his deputy.

Use of "permanent assistants" by the health insurance doctor is subject to supervision and regulation. Ordinarily

[16] Section 16 of the Road Traffic Act of 1934 provides special fees for emergency treatment rendered to persons involved in traffic accidents. There is no distinction made between the treatment of insured and uninsured persons. Emergency service to a victim of a "road accident," does not, therefore, call for payment out of the insurance funds.

[17] See pp. 33*ff*. for full discussion of payment of health insurance doctor and see pp. 39*ff*. A worker whose employment requires him to travel constantly may apply for a special voucher which entitles him to obtain medical benefit in each place he resides on a temporary residence basis.

permanent assistants may not be employed to treat insured patients without permission of the Insurance Committee. Refusal of permission, however, must be reasonable and requires preliminary consultation with the Panel Committee (the committee set up in each county to represent the health insurance doctors). If the Panel Committee, moreover, feels the consent should be granted (and only under such circumstances), the doctor may appeal to the Minister of Health for final settlement of his request. The Minister, of course, can uphold either the Panel Committee or the Insurance Committee. Thus here, as in all matters affecting the actual practice of medicine, a purely medical committee must be consulted before an administrative judgment is permitted.

The doctor who employs an assistant must remain in effective charge of all his patients. He may not in effect divide his practice with an assistant so that the latter is really responsible for part of the practice. One of the usual grounds for withholding permission to have an assistant is that there is no reasonable chance that the insurance doctor will be able to give personally the service which he is obligated to attend to in person.

The committee may consent to more than one assistant only with the approval of the Minister.

The doctor must arrange reasonable office hours to the satisfaction of his Insurance Committee and, of course, must visit and attend his patients at their homes when their condition requires it. He must also provide an adequately equipped office, as well as suitable waiting-room accommodations.

Insurance Committees differ considerably in the amount of responsibility they have taken about the condition of "surgeries" (offices) and waiting rooms. Thus, for example, as late as 1935 the London Committee undertook for the first time systematic inspection of the accommodations of all the health insurance physicians, although London, like all large cities, has so many of every grade of practitioner that admittedly there is more real need of inspection of "surgeries" than elsewhere. For some years London has required inspection before granting permission to employ a permanent assistant, and recently it has been ruled that survey of accommodations and equipment be made on each addition to the medical panel.

Birmingham, in contrast to London, began a supervisory program in the second year of the operation of the Act (1913), repeated it in 1922 and again in 1936. The entire area had been covered at least once, and all those sections deemed in need of check-up, at least twice. In addition, a special inspection has been made when a new doctor added his name to the panel and when a practitioner changed quarters. The Devonshire Committee made one complete survey five or six years ago.

In all cases the inspecting is done by the Insurance Committee with the cooperation of the Panel Committee of doctors. Usually there is a delegation of two, one doctor and one layman.[18]

"Own Arrangements"

With the approval of the Insurance Committee, insured members always have been permitted two alternatives to the customary family doctor service. One, called "collective own arrangements," is the substitution of treatment through an approved institution which furnishes medical care for its own membership (the society then receives an allotment from the committee); the other, termed "private own arrangements," is the substitution of treatment through a non-insurance doctor, toward the cost of which an annual allotment is paid.

Less than 1 per cent of those eligible to medical benefit are provided for through "collective own arrangements." Only societies that were functioning in 1911 are approved for this

[18] Far more important than committee inspection, however, in the standardization of the doctor's equipment for serving his insurance patients has been the effective cooperative effort of the British Medical Association. See pp. 76*ff*.

purpose. An increase in this substitute method is therefore impossible, except through enlarged membership of existing organizations; the membership, however, has for some time remained practically stationary.[19]

The chief advantage to the insured of being permitted to receive medical benefit through a society is that it helps to maintain an institution through which, for a small annual sum, his wife and children may also receive medical service. Otherwise he usually gets nothing that he could not receive from a partnership of doctors, for service is as a rule restricted to general practitioner service. Many doctors in different parts of England, moreover, with the approval of the British Medical Association, have been offering their services on a voluntary insurance basis to wives and children of insured workers. The chief medical quarrel with the medical aid societies is over the amount of their membership charge, which they allege to be inadequate for the proper remuneration of the doctors employed by the organizations.[20]

Only one-tenth of 1 per cent of those eligible for medical benefit make their "private own arrangements." Most of this group are required to do so because they are caught under rulings that committees are empowered to make which require persons with incomes exceeding a certain figure to make direct arrangements for their medical care. A few who are not in this high-income group are permitted to be "own arrangers" because of individual special circumstances, such as long dependence on homeopathic medical treatment but no homeopath among the insurance doctors in the area. Some of the committees now refuse to permit any such "own arrangers."

[19] "Collective own arrangements" are a survival from the days when there was no Health Insurance Act. Certain "medical aid societies" had been formed in different parts of England. These organizations made a successful fight for incorporation in the insurance act of the principle that their members might elect to continue to receive medical aid through their societies.

[20] See pp. 38*ff.* for illustrations of financial arrangements between insurance committees and "collective own arrangers'" organizations, and pp. 39*ff.* for financial provision for the "own arrangers."

Additional Treatment Benefits

In supplement to the general practitioner service and free drugs, which constitute the health insurance scheme's medical benefit, most of the larger insurance societies pay in part for various additional curative services on a voluntary extra benefit basis. A sum equal to one-third of the amount spent upon medical benefit is expended for these supplementary services.

The most commonly afforded additional benefits are dental benefits and ophthalmic benefits. Nearly ten million of the fifteen and a half million insured workers belong to societies which furnish these two "extra benefits."

The bulk of the dental work done under additional benefit has been provision of artificial dentures made necessary by long and complete neglect of the teeth. In 1934, for instance, 48 per cent of those drawing upon dental benefit had teeth filled, and 62 per cent had artificial dentures supplied. Every year, however, with the educational work which has been pressed by the Ministry of Health, a slightly larger percentage of those who draw dental benefit have remedial fillings.[21]

The ophthalmic benefit consists of contributions toward the cost of eye examinations and glasses. The British Medical Association has undertaken a drive against the use by the societies of opticians rather than medical men trained in eye work. With a group of dispensing opticians, the Association has formed the National Ophthalmic Treatment Board. Lists are offered of qualified eye men and collaborating opticians. The listed physicians will not only examine eyes and prescribe glasses, but also perform such simple operative work at the specialist's office as is feasible and, if eye disorders are detected, give the worker's health insurance doctor a report on any treatment he is capable of carrying out.[22]

[21] See Ministry of Health, "Dental Benefit," Great Britain, *Statutory Rules and Orders*, 1935, no. 644.

[22] Their services, incidentally, are offered to the members of the insured worker's family and to other persons of similar economic status at the same reduced rate.

Government regulations now stipulate that societies offering ophthalmic benefit must permit the worker to choose from this list if he desires to do so, even if the society prefers other arrangements. A subcommittee of the British Medical Association, composed of full-time eye specialists, scrutinizes applications of doctors to be added to this list. Eligibility depends upon proof that the doctor either has held special hospital or clinic appointments providing training and experience in the field or has had special post-graduate work plus experience in the field, or, failing these, that he is recognized generally by medical men as having special skill and experience in eye work.

Treatment in hospitals or convalescent homes, nursing care, and contribution toward the cost of major medical or surgical appliances are the chief other additional treatment benefits given by the health insurance societies.[23]

Statistics

According to figures furnished by the British Medical Association, in the fall of 1936, there were 22,931 general practitioners in England and Wales. Of these, 20,638 were men—14,930 health insurance doctors, and 5,708 non-insurance doctors. In other words, nearly three-fourths of the men in general practice were panel doctors.

There were 2,293 women general practitioners. Only 938 (two-fifths) of these women were health insurance doctors, and 1,355 (three-fifths) were not on the panel. Approximately 15,500,000 insured persons were entitled to medical benefit—an average of about 975 insured persons per practising health insurance doctor.[24]

[23] See pp. 91*ff.* for discussion of hospital care under "voluntary" insurance schemes.

[24] Figures furnished in September 1936. See pp. 40*ff.* for average medical income from health insurance work.

IV

REMUNERATION OF HEALTH INSURANCE DOCTORS

THE British Health Insurance Act does not prescribe any particular method of paying the health insurance doctors. From the premiums or contributions which are collected, the Ministry of Health sets aside a central medical pool for the payment of all the insured medical work. The amount that goes into this pool each year is calculated by multiplying the number of insured persons (as estimated by the government actuary) by a stipulated figure. This capitation fee (now 9s. per insured person per year) has been fixed by the Ministry of Health in agreement with the Insurance Acts Committee of the British Medical Association, which is accepted as a collective consulting agency for the doctors. In addition to the medical pool, a central mileage fund is established to cover the extra travelling expenses of practice in rural and semi-rural areas as compared with urban practice.[1]

A third fund set up is called the Practitioners Drugs Account (separate and distinct from the pool created for paying the pharmacists),[2] which is used to pay for the emergency

[1] See pp. 36*ff*. The original capitation fee (fixed in 1912) was 6s. 6d., plus 6d. for domiciliary treatment of tuberculosis and with a contingent additional amount up to 6d. from the drug fund, which brought the capitation fee up to 7s. 3d. War bonuses were paid in 1918 and 1919. A capitation of 11s. was paid during 1920-1921; 9s. 6d. in 1922 and 1923; in 1924 the fee was set at 9s. where it has remained, subject, however, to certain economy deductions in 1931-1935.

[2] Any pharmacist who desires to do so may join the panel of druggists in his area, by agreeing with the Insurance Committee to dispense on agreed terms. A *Drug Tariff*, i.e. a table of prices of drugs and scheduled appliances, etc., is issued by the Ministry of Health every month, and a dispensing fee is also payable in respect of each prescription.

Since August 1, 1937, the druggists collectively accept each year the Drug Pool (2s.3/4d. per person until January 1, 1936, since then 2s. 11d.) and accept responsi-

drugs and appliances which all panel doctors must dispense and to remunerate the doctors in rural areas who, because of the inaccessibility of a pharmacy, supply all drugs and appliances.

In supplement to the moneys set aside for remuneration of the physicians and for meeting their rural travelling expenses and their outlay in dispensing for insurance patients, a "special grants" fund, now amounting to about £8,000 ($40,000), is set apart. This is used in part as a subsidy fund to provide telephones, branch offices, automobiles, and to finance vacations for doctors in very sparsely settled districts. The object of the subsidy fund is, of course, to raise the standard of medical care by improving the facilities of doctors whose practice cannot yield a sufficient return. The "special grants" fund also supports scholarships for rural doctors, to enable them to take short "refresher" courses of post-graduate study.

These central pools are distributed by the Minister after advice from the Medical Distribution Committee,[3] among the various insurance areas (counties and county boroughs), and in these areas the administrative committees, the Insurance Committees, distribute the funds among the individual doctors. The amount to be distributed to the different areas is worked out with reference to their relative medical burden. The number of fixed insured residents, the average number of temporary residents, the mileage problems of the doctors, and other similar matters are considered. Representatives of the medical profession help in the advisory committee report, which is considered each year in apportioning the funds.

bility for the supply of all insurance medicaments dispensed by pharmacists. In most years since, they have received payment at a higher rate than the tariff rate. See Ministry of Health, 9th to 17th *Annual Reports* for details with reference to tariff.

[3] Appointed by the Minister and comprising the Government Actuary, representatives of the Department of Health, of Insurance Committees and of the health insurance doctors. This Committee gathers full particulars of the insured population and geographical conditions in each area. It makes a report to the Minister informing him what it deems to be a just apportionment of the total fund among the areas.

The General Medical Pool

The question of the method of paying for the health insurance work has been left in each area to the doctors themselves (legally[4] to the Insurance Committee plus approval of the Minister on advice of the Panel Committee; actually this advice always has been taken). Theoretically, there could be great diversity between counties in the way of paying the insurance doctors under the British Act. Actually, in all areas except Manchester and the adjoining district of Salford, the doctors by 1914 had elected one standard method of payment. The Salford and Manchester practitioners, in 1927 and 1928 respectively, also put this standard method into operation. Since then there has been but one system of payment in operation throughout the kingdom.[5]

The British doctors term this payment scheme which they have chosen the "capitation system," i.e. payment per capita of the insured on a doctor's list, be they sick or well. The doctors thus have elected to have their remuneration vary with the number of people for whose care they are responsible, rather than with the items of medical service they may render.

The doctor receives a government check every quarter. Its amount is determined primarily by the number of names on his medical list at the beginning of the period. Persons added during the quarter are not counted until the next quarter, but this is balanced by the rule that no credit is lost for those who go off a list in this period. A single account may be rendered to a partnership, and this is usually done.

The exact method used in distributing the area's money among the doctors is as follows:

[4] See Sec. 32 of the National Health Insurance Act, also *Medical Benefit Regulations*, Sec. 23.

[5] Kent, Manchester, and Salford in 1913 adopted a straight fee system, a few others adopted a combination of capitation and fees. See pp. 41*ff*. for exposition of Manchester experience prior to 1927 and see 1st *Annual Report* of the National Health Insurance Commission for detailed account of this stormy first year, pp. 125-48.

The Insurance Committee of the area sets aside each quarter a sum for expenses of the Panel Committee (the insurance doctors' committee) and for any exceptional services which are specially paid for,[6] such as administering anesthetics. To determine distribution of the remaining funds, units of credit are first calculated:

> 1 unit for each name on each doctor's list (at the beginning of the quarter)
>
> 4 or 5 units (or whatever number the committee has decided on) for each temporary resident treated throughout the quarter in the area. This number is determined on the basis of the relative burden of the temporary as compared to the permanent residents.

The total number of units of credit is then divided into the amount to be apportioned among the doctors. This gives the quarterly capitation figure.

The amount due the individual doctor is then computed by first multiplying this capitation figure by the number of his units of credit and then adding to this amount special fees for any special fee work he has performed for someone else's patients and subtracting any special fees which have been paid another practitioner for similar service to his patients.

ILLUSTRATION OF A QUARTERLY SETTLEMENT

One quarter of the amount ($100,000) allotted to the area for the year		$25,000
Deductions for Panel Committee		
Expenses	$100	
Anesthetics	100	
Other special payments	50	
Total Deductions		250
Net amount to be distributed		$24,750

[6] This usually includes fees for administering anesthetics and rarely for certain other special services, such as night visits and certain operations.

There are 40,000 units of credit on all doctors' lists.

There are 2,150 units on Dr. Brown's list (2,000 regular residents, 30 temporary residents treated, each counted as 5 units).

Dr. Brown has performed no emergency work for other doctors' patients. No doctor has performed emergency work for his patients.

Dr. Brown will receive for the *quarter*:

$$\$24,750 \div 40,000 \times 2,150 = \$1,328.70$$

Mileage Fund

The central mileage fund is itself divided into two parts. The so-called Main Part is used to pay allowances for ordinary rural travelling beyond two miles from the doctor's house. The Special Part is used for additional allowances for travel that entails exceptional difficulties, such as use of bad roads and footpaths, fording streams, crossing lakes, etc.

It should be emphasized that mileage payments are not intended to cover the cost of *all* travelling entailed by medical practice, but merely the excess of travel required in a rural or semi-rural practice as compared with town practice.

Computations for mileage allowances were first put on the basis of actual experience by using statistics of 750 typical practices in 1921 and 1922.[7] Periodic revision is based on similar first-hand information about the actual situation of representative insurance doctors. About a fifth of the Special Part of the mileage fund is used to pay exceptional travel allowances to doctors in certain very sparsely settled areas (which do not offer a remunerative practice) as a form of subsidy to ensure the likelihood of a resident doctor.

The committees in the different insurance areas may not mix their Special Part grant with their Main Part allotment.

[7] As a result of study of these statistics roughly 6,000,000 miles was estimated as the total rural or semi-rural travel in England and Wales, 1s. a mile was allotted and 9½d. per patient was deducted as representing the amount expended by town and country doctors alike in reaching patients in need of visits. The residue constituted $1,250,000, of which $1,025,000 was allocated to ordinary allowances.

The relative amounts of these shares vary, of course, with the travelling problems of the counties. Thus Westmorland's Special Part grant is almost as large as its Main Part. Essex, on the other hand, has few spots difficult of access, and accordingly, its Main Part allotment is almost twenty times as great as its Special share. Devon's Special mileage is about one-third of its Main mileage fund. Middlesex and Nottinghamshire need only a Main Part allowance.

In each area the individual Insurance Committee, in collaboration with the Panel Committee and with the approval of the Minister of Health, decides on a "scheme" for dividing up their area's share of the Main Part and the Special Part. Actually, the Main Part is divided in most insurance areas in much the same way. The usual system of mileage allowance is on a capitation basis, just as is the doctor's remuneration for medical service. Mileage is paid for each person on the medical list who lives at a distance, whether or not the doctor in fact is called upon to visit him in the particular three months' period covered by the payment. In consequence, mileage payments vary with the distance of *potential* travel to *potential* patients, rather than with the actual miles the doctor may drive in any particular quarter. Units are calculated as follows: First, one unit is allowed the doctor for each patient in respect to each mile (or fraction thereof) in excess of two in the distance between the patient's house and the doctor's office. Thus, for example, the practitioner gets one unit for each patient who lives between two and three miles distant, two units for those who live between three and four miles distant, and so on. If several members of the household are insured, the doctor is allowed the appropriate units for each. Second, the doctor's practice is classified from a travel standpoint as (1) easy, (2) ordinary, or (3) difficult. Deductions are made from the computed units for an "easy" practice. Additions are made to the computed units for a "difficult" practice.

There is considerable diversity in the methods used in dividing up the Special Part.[8]

The Practitioner's Drugs Account

Each insurance area distributes its share of the Practitioner's Drugs Account on the basis of either capitation or "payment per prescription dispensed in accordance with the *Drug Tariff*." If the latter type of payment is adopted, the amount paid includes (1) actual cost of drug or dressing, plus (2) the dispensing fee. (The *Drug Tariff* is the official list of prices, revised monthly, on the basis of which the Insurance Committee evaluates the drugs.)

The capitation method of payment is the one generally adopted and is supplemented by special payment for specified exceptionally expensive drugs. The drugs capitation fee is uniform for all areas and, like the capitation amount for medical services, is negotiated centrally by the Insurance Acts Committee of the British Medical Association. The 1936 figure was set at 2s. 3d. (56 cents) per head if all drugs are supplied, and 1s. 3d. (31 cents) per 100 persons on the list if only emergency drugs and appliances are dispensed by the doctor.

Collective "Own Arrangements"

As explained in the previous section, about 1 per cent of the insured persons receive their medical care from Approved Medical Aid Societies. The doctors furnished by these societies are salaried employees of these organizations. The Insurance Committee, on receiving a certified report of the society's expenditures, pays the society either the percentage of its expenditures which may be properly allocated to medical care of the insured membership or the amount on hand in the committee's hands for such care, whichever is smaller. The percentage properly attributable to the insured depends upon the ratio of total expenditures for medical

[8] Interesting examples are found in the mileage "schemes" of Devonshire and Cumberland.

service represented by the ratio of the insured members to the total membership. It is agreed that the cost of medical service for insured members should be weighted, since the non-insured members of an Approved Medical Aid Society are required to pass a medical examination, and therefore are selected risks. Frequently the committee counts two insured members as equivalent to three uninsured members.

ILLUSTRATION

An Approved Society (in Derbyshire) with a total membership of 6,800 had 4,400 "insured members" in 1935.

Figure 1.	The cost of Society's medical care which may be properly allocated to insured members	$12,250
Figure 2.	The amount allocated by the Insurance Committee to furnishing 4,400 persons medical care	$11,700

Since Figure 2 was less than that actually expended (Figure 1), the society received the full $11,700. Had Figure 1 (the amount spent) been less than Figure 2 (what the Committee *could* spend), the smaller would have been paid.[9]

Individual "Own Arrangers"

A sum equal to the combined annual amount due from the insurance pool for medical care for the "own arrangers" is placed each year by the committee in a separate account. At the end of the year, the bills incurred for medical care, calculated in accordance with a scale of fees fixed by the committee, are presented by all the "own arrangers." The committee may disallow items which they deem excessive (the insured is allowed to present his case for full allowance). The "own arrangers" then receive their proportionate share from the account. If there has been but little sickness and there is enough to cover all the bills, those who present bills may find them paid in full. These bills, however, are cal-

[9] In the case of this institution, the committee counted two insured members as equivalent to three uninsured members.

culated in accordance with fixed fees and may be far less than the actual bills of the physician. If there has been a great deal of illness and demands are heavy, the "own arrangers" may get back less than their presented bills and a still smaller fraction of what they have paid to their physicians.

Altogether, 13s. per insured person is guaranteed for meeting the costs of medical benefit. The total health insurance income of all the doctors practising under the scheme in 1935 in England was £6,922,000 ($34,610,000).[10] Of this total, £6,520,000 ($32,600,000) was for medical services, £204,000 ($1,020,000) for mileage, £191,000 ($955,000) for medicines dispensed (10 per cent estimated to be profit), and nearly £7,000 ($35,000) in "special grants" to rural doctors. The latter included a few subsidy grants of as much as £275 ($1,375) to doctors in remote and inaccessible districts and also grants which enabled forty-five doctors to take postgraduate summer courses. The average health insurance income per insurance doctor, counting the first two and fourth items and 10 per cent of the third, was approximately £426 ($2,150).[11]

Foreign income unrelated to foreign wage levels, when stated in terms of American currency, is relatively meaningless. It is useful, therefore, to compare this medical income with the maximum earning capacity of (1) the skilled British worker in the engineering trades, (2) the unskilled British laborer in the engineering trades, and (3) the agricultural laborer. The maximum annual earning capacity of such a skilled worker, assuming fifty weeks of employment, was £155 ($775); that of the unskilled engineering laborer was £105 ($525); that of the agricultural laborer £80 16/ ($404),

[10] There were the following additional payments from the Health Insurance Funds "for medical service (including drugs)" to insured persons. The sum of £63,700 ($318,500) was paid to approved institutions on account of insured persons and £7,900 ($39,500) to "private own arrangers" for medical service from non-panel doctors.

[11] The number of insured eligible to medical benefit in 1935 was approximately 15,000,000. Nine shillings was the capitation fee, subject, however, to an economy deduction for the first half year of 1935.

assuming fifty-two weeks of employment.[12] The average medical income from health insurance thus equalled almost two and four-fifths times the average maximum earnings of the skilled engineering worker, a little more than four times that of the laborer in the same industry, and more than five times that of the agricultural laborer.

It should not be overlooked that the health insurance income of the British doctor represents merely his income on account of the insured wage earners. Services to the families of these wage earners, as well as those rendered the better-paid non-insured persons, of course mean additional receipts. Such services are not all on a fee basis. Much of the treatment of families of the insured is now on an insurance basis through a supplementary voluntary insurance organized by the doctors and sponsored by the British Medical Association pending the addition of family benefits to the regular insurance scheme. A small weekly contribution is collected—usually 6d. (12 cents)—and the doctors render the family service on a capitation basis just as they do the regular insurance service to the workers. The usual amount of supplementary income collected for non-insurance work was estimated in all parts of England as sufficient to make total medical income about double the insurance receipts in the case of a doctor with a practice confined to insured workers' families and proportionately more with an increasing ratio of practice among the better-to-do.

The "Manchester System" of Payment

The so-called Manchester System, i.e. payment by fee for each unit of medical service rendered, as has been stated, was not retained after 1914 by any of the small groups of counties which originally adopted it, except Manchester and the neighboring Salford district. The Salford and Manchester

[12] See *Great Britain, 79th Statistical Abstract for the United Kingdom*, 1936, cmd. 5144, p. 139, for wage data.

doctors gave it up at the end of 1926 and 1927 respectively, after fifteen and sixteen years' experience with it.

Doctors under the attendance system were required to keep service records of each patient treated and each item of service rendered. Patients were allowed, of course, to shift freely from doctor to doctor at will. The doctors' service records were presented each quarter to the Panel Committee of doctors, and this committee determined the actual amount to be paid the individual practitioners in accordance with a schedule agreed to by the Insurance Committee and the Panel Committee.

There were seven types of service recorded, and according to the revised schedule they were weighted as follows:[13]

1. Office visit—1 unit.
2. Ordinary house visit (daytime visit during usual calling hours of the M.D.)—1.5 units.
3. Special visit (urgent visit at other than usual calling hours of the M.D.)—4 units.
4. Night visits (between 9:00 p.m. and 9:00 a.m. in response to call received during those hours)—4 units.
5. Surgical operation requiring local anesthetic—4 units.
6. Administration of general anesthetic—14 units.
7. Setting of fracture or reduction of dislocation—14 units.

The total number of units for all the doctors was then divided into the sum allotted to the area for payment of the

[13] Until 1924, instead of being computed as units, a money equivalent was used as follows:

Office visit—3s.
Ordinary house visit—4s.
Special visit—5s.
Night visit—10s.
Surgical operation requiring local anesthetic—10s. 6d.
Administration of general anesthetic—£2 2s.
Setting of fracture or reduction of dislocation—£2 2s.

area's insured medical service. This furnished the fee per unit of service rendered.

At first the doctors' records, in the absence of allegation of fraud, were accepted as presented. In the course of time, however, such dissatisfaction was expressed by many doctors at the relatively high number of visits of certain practitioners, that the physicians directed the Panel Committee to work out some form of equitable regulation of accounts. This regulation became gradually standardized along the following lines. From *all* the accounts the average unit credit per patient treated was computed. Then the average per patient for each doctor was computed. In the case of each doctor whose individual average exceeded the general average, the Panel Committee made inquiry to determine whether special circumstances justified the difference. Before any account was actually reduced, notice was sent to the doctor concerned asking him either to appear or to submit a written statement defending his charges. Since, in best of faith, judgment of medical men differs as to the number of visits necessary during a given illness, bitter feeling was often engendered by the Panel Committee's reduction of accounts. In Salford the committee was pushed into still further standardization. An arbitrary standard of proper average attendance was fixed by the Panel Committee, and all doctors' claims that exceeded this standard were automatically reduced to conform to this average. The practitioner then was required to establish his right to an additional crediting, should he insist upon such.

Meantime, despite the Panel Committee's supervision, the number of total items of service mounted steadily year by year. General medical opinion has it that this was caused by the small minority of unscrupulous physicians, the inevitable chiselers who are found in all walks of life. Unable, because of the Panel Committee's control, to run up bills indefinitely in the individual illness case, apparently they simply spread their claims over a greater number of patients. Sometimes this was done by sheer prevarication about pretended visits, operations, etc.—so clumsily done on occasion that the fraud was obvious to the Panel Committee. Sometimes doctors forgot the number of days in a month so that, as one committee secretary put it, "When they said they visited patients on February 30 and 31 and June and September 31, they were apt to be doubted when they said that they stayed away from golf to visit people on Sundays."

More often, however, inflation of the number of visits was of a character that was impossible to control. Workers were encouraged by the little group of trouble-making doctors to exaggerate their trifling ailments, urged to come in for unnecessary prescriptions of palatable tonics, and otherwise induced to collaborate in the swelling of the items of the service rendered.

The result, of course, was that with a fixed sum to divide and a constantly increasing number of units of service to be paid for, the payment per item of service became steadily less with each successive year.

The following statistics of the final six years' experience under the fee system show the mounting number of annual credits *disallowed* by the Panel Committee and, despite this, the steadily increasing number of credits allowed and, in consequence, the *diminishing cash value* of the unit of service rendered.

Manchester Insurance Committee

Statistics Relating to Accounts of Practitioners[14] Under the Attendance System of Medical Benefit for the Years 1922-1927

Year	1922	1923	1924	1925	1926	1927
Amount* for medical care allowed the area per insured person per year	$2.38	$2.38	$2.25	$2.25	$2.25	$2.25
Units disallowed by Panel Committee	85,974†	109,515†	106,772	139,250	173,384	205,991
Total units in accounts allowed	1,402,318	1,530,203	1,662,734	1,813,197	1,944,654	2,168,361
Average cash value* per unit credited	$.52	$.46	$.40	$.37	$.36	$.32

*Counting a shilling at twenty-five cents.

†Computed approximately. Disallowances actually computed in *cash* these years.

[14] Furnished by Mr. Dewhurst, assistant secretary of the Manchester Insurance Committee in June 1936.

Reduction in value of the unit of service inflicted hardship upon the great majority of men, who were conscientious and scrupulous about their accounts, to the advantage of the small group whose inflated accounts were causing the trouble. Similarly, the rule which had been made in the interest of checking frauds, that no doctor without proof could have credit for more than "average" services, worked such an imposition upon the time and energy of the reliable busy man that he often forfeited a just claim rather than pursue his remedy.

The final determination of the doctors to abandon the fee system in favor of the capitation arrangement long in operation in the rest of Britain was concurred in by the overwhelming majority of the practitioners in the area. Nevertheless, there are still to be found a few dissenters among the older medical men who resisted the change and still feel that further adjustments should have been tried before giving up payment "according to unit of service rendered." There seems to be no dissent, however, to the following explanation of the surrender of the fee system: control was required because it was impossible for all doctors to trust each other, yet control, even when in the hands of medical men, pushed inevitably toward standardization of average amounts for specific medical cases; such control worked hardship to the conscientious and scrupulous men and yet was incapable of preventing imposition by those disposed to take advantage.

The general sentiment relative to the change to capitation seems to be profound relief that bickering over accounts is ended and that bookkeeping is transferred to the Insurance Committee, which pays the doctor according to a uniform system that plays no favorites and leaves the doctor to practise his profession in peace.[15]

[15] In the *Report of the Royal Com. on National Health Insurance*, 1924, Appendix C at p. 91, it was mentioned that some of the Insurance Committees wished to put the Manchester system in effect (to obviate the work of keeping Medical Lists, etc.), but the opposition of the medical profession was so great that capitation was retained.

V

THE WORK INVOLVED IN HEALTH INSURANCE PRACTICE

Medical Work

IN turning to the actual work done by the British health insurance doctor, it must be borne in mind that invalidity and health insurance are merged in Britain. The doctor therefore serves in a threefold capacity. He is, to the extent that the services of a general practitioner are needed, doctor (1) for those who are normally at work and are temporarily ill or injured (including those suffering industrial injury), and (2) for those who are permanently invalided and drawing invalidity pensions, and (3) for those who are more than sixty-five years of age and are drawing old-age pensions through the Insurance Funds.[1] More than 90 per cent of the persons he serves are of working age.

Official figures of the actual average units of medical work per insured person occasioned by health insurance practice were not made public until the summer of 1937. At that time evidence based upon the experience of 10 per cent of the British Medical Association membership in each area was presented to a governmental Court of Enquiry.[2] This placed on record the following statistics:

[1] This does not include, of course, persons drawing the gratuitous pension paid on proof of need to persons over seventy, but merely those who receive their pensions as a right under the insurance system. cf. Chap. II, n. 4, *supra*; in re special payments for emergency care in traffic cases.

[2] This court was appointed to consider whether the capitation of 9s. should be altered. The British Medical Association asked for an increase of 12s. 6d. The Ministry of Health asked for a reduction to 8s. The court's decision was that the 9s. level should remain unchanged and this was accepted by the parties. See full account of this evidence in *British Medical Journal*, issues of May 29, and June 12, 1937 and see n. 3, *infra.*

Number of insured persons involved in the sampling study	1,590,207
Office consultations	6,054,268
House calls	2,115,992
Average consultations per insured	3.80
Average house calls per insured	1.33
Average items (i.e. either consultations or house calls per insured)	5.13

For the seven years 1930 to 1936, inclusive, the average number of consultations was 3.705, the average number of house calls was 1.315, thus constituting an average of 5.02 items per insured. This service, it must be remembered, is for the care of both temporary illness (including industrial injury) and invalidity.

Specific records of individual practitioners with rural and town practices selected at random serve to illustrate the variation in percentage of *potential* patients who became *real* patients, i.e. the fraction of the medical list actually seen in health insurance work. Thus, a rural practitioner in Lincolnshire saw the following percentages of his patients in the four quarters in 1935: 30, 32, 23, and 35 per cent. A man in a neighboring town saw a much heavier percentage, namely, 55, 59, 59, and 58 per cent. A small-town practitioner in Nottingham saw 52 per cent of his list during the year, an average of 5.4 times each. In the previous year he saw 49 per cent of his list an average of 5.3 times. Another man in a neighboring town saw nearly 60 per cent of his list. Undoubtedly the percentage of the list seen by the town practitioner with his accessible "surgery" is heavier than that of his colleague in rural practice, and with equal assurance it may be said that the town man sees relatively more patients at his office and fewer at their homes.

The assortment of services rendered by the health insurance physician includes, of course, giving drug prescriptions (for only with a prescription from his doctor may the worker secure his medicaments without cost), giving certificates of incapacity to enable his patients to get their cash benefit, as well as actual physical check-overs at his office or visits at the patients' homes.[3]

No official figures are available to indicate the distribution of units among the various types of service—beyond the previously mentioned distribution between office and home visits.

Scattered sources of information from well known individual practitioners[4] suggest that even in the worst months of the year not more than a fifth of the patients seen in insurance practice require a real examination or "going over." The other 80 per cent come in (or are visited) for prescriptions, for certificates, or for a "follow up" check of a condition understood from an earlier complete examination.

Paper Work: Certification

As stated in Chapter II,[5] one of the obligations accepted by the health insurance doctor is that of "certification." On request he must give a "certificate of incapacity" to a patient who, in his medical judgment, should not be at work. The patient then presents the certificate to his health insurance society as evidence that his condition calls for the weekly cash payment. And the doctor must issue a "final certificate" when he believes his patient is once more fit to work. Books of certificates are issued to the doctors by the Insurance Committee, stamped with the name and address of the doctor.

Decision to pay the cash sickness (or disability) benefit is made, in the last analysis, by the society—ordinarily, however, payment is made on the basis of the doctor's certificate. In fact, as it is put by the *Medical Insurance Practice* book issued by the British Medical Association, "Every certificate is, in effect, a check drawn on the Benefit Funds of Approved

[3] The Ministry of Health in a sampling study in 1936 of 660 practices, found only 3.66 units of service per card. See *British Med. Journ.*, Supp. 1937, p. 320.

[4] Each of whose medical lists totalled two thousand or more.

[5] p. 15 *supra.*

Societies and even slight laxity in the issue of certificates may produce serious financial results."[6]

Seven types of certificate are used by the doctor in insurance practice: a death certificate, the first and final certificates already mentioned, three kinds of "intermediate certificate," and a so-called "voluntary certificate."

The ordinary "intermediate certificate" is really a renewal certificate signed every week during the continuance of incapacity. In cases where a long recovery period is to be expected and where weekly examination and treatment is unnecessary, a "special intermediate certificate" may be issued at intervals of not more than four weeks. Further, a "convalescent intermediate certificate," covering a two weeks' period away from home for convalescent purposes, may be used when a patient has had at least four weeks' treatment. If he goes away to recuperate after a shorter period, he must get weekly certificates from a doctor in the locality he visits.

The first and intermediate certificates are due "on request" of the worker if the doctor believes them justified. The final, whether requested or not, must be given when the doctor believes his patient is ready for work again. These must be signed either at the time of actual examination or within twenty-four hours of this time. The "voluntary certificate," on the other hand, may be issued retroactively, and is intended to cover cases where a worker who is actually incapacitated and entitled to cash benefit has not asked his doctor for a certificate at the proper time. A special charge may be made for a certificate of this sort, though most doctors do not take advantage of the privilege.

Except in cases in which a statement of the disease or disability suffered would be a hazard to the health of the patient, the certificate must indicate the specific ailment which causes the incapacity. Obviously, it is frequently impossible,

[6] Harris and Sack, *Medical Insurance Practice*, p. 104.

because of difficulties of diagnosis, to state more than general symptoms in many first and even intermediate certificates. The physician must, however, be as precise as possible and, with the continuance of incapacity, is expected ordinarily to be able to give increasingly exact information.

A so-called "vague certificate" is given when the physician deems it unwise that the patient should know his actual condition. In such cases, a real report is sent to the Regional Medical Officer,[7] together with an explanation of the need for withholding the real information.

In giving certificates of incapacity, the doctor is passing judgment upon a condition of his patient described by the law in terms that are susceptible of various interpretations.

"Incapable of work" literally construed would, of course, mean unable to perform *any* work. It is tacitly agreed, however, that the phrase means ordinarily, i.e. in cases of temporary sickness as contrasted with invalidity cases, *unable to follow his ordinary occupation*. It is deemed unreasonable and not within the contemplation of the insurance law that a worker whose physical condition temporarily keeps him from his regular work should in the interim be expected to find some unusual employment which he might physically be fit to perform.

Occasionally a patient is consciously or unconsciously a malingerer, who resents ever being sent back to work. Occasionally the doctor is not certain that his judgment of his patient's fitness for work is accurate. In either case, the doctor may, if he chooses, certify incapacity and then refer the case to the Regional Medical Officer for second opinion. Apparently the extent to which the privilege is used varies greatly in different areas and also among the individual practitioners. In a few instances doctors stated that they never "referred," feeling that the insurance society usually told the patient what the doctor had done, thus causing more trouble than if the doctor met the issue squarely with his

[7] See pp. 53*ff.* for discussion of the function of the Regional Medical Officer.

patient. Some practitioners, in areas where no distrust of the society's discretion was complained of, expressed the view that a doctor as a matter of personal pride should handle his own difficult patients without relying on the Regional Medical Officer.[8] Other doctors, however, felt that in the occasional uncomfortable situations (the doctors stress their infrequency), it was a welcome respite to turn the matter over to the government physician.

In cases of permanent impairment, when it becomes obvious that the patient will never be able to resume his previous work, the definition of incapacity as "unable to follow his ordinary occupation" no longer serves. "Incapable of work" must then be construed by the doctor as meaning inability to perform any suitable kind of remunerative work. The society from which he obtains his cash benefit then allows him benefit for a reasonable period of adjustment, during which he can equip himself for some substitute occupation. The member can appeal to the Ministry if he deems that he is unreasonably cut off from benefit under these circumstances.

Sometimes the doctor is notified that a patient belongs to a society which gives "additional treatment benefits" affording therapy which the insurance doctor has advised. In such cases, if the patient requests it, he must recommend him for this benefit on the next regular certificate which the doctor has occasion to sign for him.

In all cases, the doctor on request must give a written statement recommending ophthalmic treatment to any of his patients whom he deems to be in need of such specialist service.

"Records of Illness"

The insurance doctor is required to keep a medical record for each of his patients. The Regional Medical Officer must be given access to these records and has the right to sup-

[8] See pp. 56*ff*.

plementary information. Office visits and house calls must be recorded, and reasonably full clinical notes of the ailments treated must be included. It is not expected, however, that the doctor will record details of every trivial symptom or disturbance.

The patient's file is sent to the Insurance Committee for forwarding to the appropriate physician in case of change of doctor. The record is handled as a confidential document by a responsible and specially deputized officer of the committee.

The records are supposed to serve a twofold function: (1) to give each attending doctor an account of his patient's previous medical history; (2) to give the administrative agency acting through the Regional Medical Officer material on the basis of which prevention of illness can be planned. Incidentally, the record of office visits and house calls enables the Regional Medical staff to check up the occasional practitioner who gives his patients conspicuously inadequate attention.

Tuberculosis Cases

Special forms must be used to advise the Tuberculosis Officer in each area of cases of both diagnosed and also suspected tuberculosis. Terms of service provide that the same reporting may be required, at the discretion of the Minister of Health, for other diseases.

If the Tuberculosis Officer decides that a patient needs treatment at a public dispensary or a sanatorium, and the patient accepts such treatment, the obligations for certificate of incapacity, as well as for medical care, are of course transferred to the Tuberculosis Officer for the period of such treatment. If merely general supervision is decided upon and the patient remains in his own doctor's hands, the practitioner is required to report periodically to the Tuberculosis Officer the progress of each case.

VI

THE REGIONAL MEDICAL STAFF

SINCE 1920 there has been a network of medical administrative officers in connection with health insurance.[1] This system was set up and has been retained largely in response to demand from the health insurance doctors for medical referees to check certification cases. The Minister of Health is empowered, however, to require these officers to perform any duties that he deems suitable.

England is divided into four divisions (Wales comprises an additional region), each of the regions being presided over by a Divisional Medical Officer. Leeds has the divisional office for the northeastern division, Manchester the divisional office for northwestern England, while both the southern divisional headquarters are in London.

Each Divisional Officer has in turn a staff of Regional Medical Officers (commonly referred to as R.M.O.'s) and deputies attached to regional offices scattered throughout the division. The full-time staffs are supplemented by additional part-time medical referees drawn from the practising (and occasionally retired) medical men in the community. These part-time men do nothing but check certification. They are used during the "peak" periods of illness claims, when a great bunching of references occurs. They were added to the regular staff because it was the most economical way of meeting the demands of these periods, but many officers believe that the occasional administrative work of these practitioners has brought a most beneficial by-product: a more harmonious relationship between the medical men in different localities

[1] A scheme of medical referees prepared by the Insurance Commission in 1914 with supporting grant voted in July was interrupted because of the War. See National Health Insurance Commission, 2nd *Annual Report*, 1914, cmd. 1422, p. 44.

and the permanent medical staff responsible for the administration of the health insurance scheme. By permeating the practising profession with men cognizant of and sympathetic with the problems of the Regional Medical Staff, cooperation of the panel practitioners and the administrative officers has been promoted.

More than half the time of the medical officers is given to medical examinations. Most of these examinations are made on request of the health insurance societies for advisory opinions about eligibility of insured persons for cash benefit because of incapacity for work.

These advisory opinions do not control the judgment of the attending doctor. Even if a worker is declared fit by his referee, the practitioner may continue to certify his patient as incapacitated for work. Further, if the society decides to be guided by the referee's judgment and refuses to continue benefit, the worker may appeal to the Ministry. In actual practice in the vast majority of cases the opinion of the referee is accepted as final by both the worker and his society.

Not all the requests for referee examination, or "references" as they are called, are based on actual suspicion of malingering. Some are random selections from the lists of those on benefit for an extended period. The latter are intended to serve constant notice to the insured workers that they must not exaggerate their ills. It is also considered wholesome that the occasional careless doctor should know that his reputation will suffer if too many of his patients are sent to the referee. Most of the referred persons are seen at the regional office. If the condition of the patient is alleged to require it, however, the medical officer goes to the worker's home.

There has been a marked and constant decrease in the number of annual "references" in the last four years. This is widely attributed to improved relations between the medical profession and the health insurance societies, based on a growing confidence that the doctors in general are scrupulously fulfilling their certification responsibilities. It is also attributed in part to "good times," i.e. the fact that this has been a period of increasing prosperity with higher wage opportunities that promote especially prompt return to work, even by persons who would be better off for another day at home.

The total number of references[2] for both men and women was reduced by roughly 25 per cent in the four years 1932-1935 inclusive. There was a reduction of almost 29 per cent in the references for insured men, and approximately 23 per cent in the references for women. The total number of persons "notified" in 1935 was the equivalent of only 2.7 per cent of the persons insured for cash as well as medical benefit. Quite a number of both the men and the women in this small group of "notified" persons (i.e. those who are asked to see the Regional Officer) either at once withdraw their request for benefit or fail to appear for examination. In the opinion of the medical referees, this is a result of fear of a "strange" doctor as often as it is an indication that they have been exaggerating their ills.

Certification difficulties are felt to be more pronounced with insured women than with the male workers, as is obvious from the fact that women, although constituting only 35 per cent of the insured population, account for 64 per cent of the cases "notified." About 22 per cent of the men examined (who comprise just about 1½ per cent of the men insured) are stated to be fit for work, while nearly 32 per cent of the women (who comprise about 2 per cent of the women insured) are declared to be able to return to their jobs.

This difference between male and female workers is attributed by the Insurance Committee officials and by the Regional Medical Staff mainly to two factors: first, that most women are in industries of such low wage standards that the margin between the amount of their benefit and their

[2] Information furnished by the Ministry of Health, London, August 1936.

wage is slight; they have less incentive than men to get back on the job promptly, and so are more prompted to "malinger." This lack of incentive is pronounced in the case of married women who, at the end of the day's work, have the care of their homes and families awaiting them; second (and this is emphasized as the chief factor), that because of very low wages and consequent standard of living, many women are constantly in such poor physical condition that it is difficult for the attending doctor ever to certify them as "fit and capable" of work after they have been drawing benefit because of an acute illness.

In health insurance, as in everything else, low standards of wages and living breed special difficulties.

In addition to making medical examinations for decisions as to incapacity, the R.M.O. (Regional Medical Officer) is available to the health insurance doctor for consultation purposes. These consultations are for purely medical purposes. The number of requests for such consultations is still very small, although growing slowly from year to year.

The R.M.O. also enables the Ministry of Health to reach and aid persons found suffering from silicosis and from blindness. He conducts a certain number of examinations as a courtesy to the Ministry of Labor to determine whether workers are in suitable physical condition to enter the industrial training centers conducted in connection with unemployment insurance. He also inspects, for both insurance and non-insurance doctors, the records required under the Dangerous Drugs Regulations.

The R.M.O. puts in about one-third of his time visiting the individual practitioners for administrative check-up on the keeping of records, the prescription of drugs, and the certification work. In this visiting work the effective R.M.O. is more an educator than a mere supervisor. He is, as one doctor to another, interpreting the insurance scheme to the practitioner, teaching him the importance of careful, economical prescription, persuading him to greater exactness in certification, helping him to simpler, more efficient record-keeping—in short, constantly doing what many of them call the "spade work." Not only medical skill and medical reputation, but also tact and a thorough understanding of *homo sapiens* are needed by the Regional Medical Officers.

There is evidence in the functioning of the scheme that most of these experienced practitioners (the R.M.O. must have had ten years in health insurance practice) have been well selected.

In 1935, the Regional Medical Staff in England made 232,457 incapacity examinations and 1,047 examinations for "consultation" purposes. In addition, the staff examined 23,365 unemployed persons to determine the question of physical fitness for entering Industrial Training Centers, 539 men and women suspected of suffering from silicosis and other special occupational diseases, and 225 persons applying for government pensions granted to the blind.[3]

[3] See 17th *Annual Report* of the Ministry of Health, 1935-1936, p. 186.

VII

MEDICAL ATTITUDE TOWARD CERTIFICATION RESPONSIBILITIES

THE British health insurance doctor is not disturbed today by his certification responsibilities. He accepts them as a basic part of his medical duties, exactly as he does the prescription of drugs for his patients. Indeed, the last three years have seen such an intensification of the administrative drive against the "over-prescribing" of drugs by the health insurance doctors, that the average doctor, for the time being at least, is much more kindly disposed toward his certification responsibilities than toward his prescription duties. This was not always the case. Although most doctors seem to have forgotten it, many of them were so annoyed by the accusation made in 1931 by the Friendly Societies (i.e. the societies paying the cash benefits to incapacitated workers) that there was widespread laxity on the part of the doctors in their certification, that there was resentment about the whole question. The special governmental investigation prompted by this accusation resulted in a public report that was deemed by the medical profession to clear it of the charges. Since that time, relations between the societies and the British medical profession have become so harmonious that the episode is largely buried in oblivion. Indeed, the practitioner is actually surprised at the suggestion that he may be annoyed by having to "certify."

In every part of England, responses were the same to queries about certification obligations. To wit: "All paper work was a nuisance but this was a necessary obligation, and through convenient forms, had been reduced to a minimum, and payment for this certification work was counted on in computing the capitation fee." As to the assumption that the responsibility itself created a real problem, this simply was not a fact. The average run of workers did not put their doctor "on the spot" about certificates that would put them either on or off cash benefit. This occurred in only a small number of cases. Most workers, in fact, wanted to get back to work before they were really fit to do so. As to the occasional "loafer" who never wanted to work, he was a nuisance, of course. But so was the man who was always wanting a tonic when the doctor knew he didn't need it. The doctor had to be man enough to handle his occasional "do-nothing" and his "medicine bottle addicts." And with the former group, at least the doctor who did not want to meet the issue, could always put the case before the Regional Medical Officer.

The remarks of one[1] of the British Medical Association's secretaries on the subject are pointed and bear repeating. "There is no 'certification problem' " said he, "in the sense in which you are discussing it. The idea that there is, is based on the wholly false assumption that most patients are struggling to stay home and draw benefit. That, of course, is utter nonsense. It's a relic of the same kind of thinking that pictured workmen's compensation as a social calamity on the ground that injured workers would become malingerers who would never want to work again."

Everyone connected with health insurance frankly concedes, to be sure, that under Britain's democratic system the attending practitioner tends to be more lenient in his certification than he would be if on a salary and responsible only to the state, without direct personal relationship to the patient. Panel doctors generally agreed, moreover, that the medical referee is necessary both to bring into line that small percentage of lazy workers found everywhere and also to stiffen the occasional doctor who lacks moral courage. It is

[1] Dr. Charles Hill, Health Insurance Secretary of the British Medical Association, in an interview June 1936.

confirmed medical opinion in Britain that a free system, open to all doctors, in the nature of things calls for supervision.

Queries as to the desirability of turning over the whole question of certification to the Regional Medical Officers brought the same reply from practitioners everywhere. It was an obviously unworkable suggestion. The doctors would resent it as interference with their patients and it would be administratively impossible as a practical matter. No one but the attending doctors could know the condition of the sick workers. It would mean duplicating the whole practising profession by administrative officers who would personally examine all insured patients, and even then the attending doctor would have to be consulted in many cases.

These opinions seem to represent the whole practising profession and not just the particular men in the various areas who expressed them, as is evidenced by the following statement.[2]

"For your information, I have since 1924 been connected with the central work of the Insurance Acts Committee. For the last four years, I was chairman of the Panel Conference[3] and during the whole of that time no suggestion has even been made by any Panel Committee, or by any individual insurance practitioner, that certification should be taken from the hands of the general practitioner. It is scarcely conceivable that there is any real suggestion for this procedure without its having been brought at some time or another to the notice of the official body. I have no doubt that it is possible to find men who are in insurance practice who would say that they did not wish to continue to do the certification, but I very much doubt whether there is a single Panel Com-

[2] Contained in a letter (June 11, 1936) sent to the author by Dr. Herbert C. Jonas, chairman of the Insurance Acts Committee of the British Medical Association, the central committee of the British Medical Association, which is accepted by the British Government as the official representative agency of the practising health insurance doctors. See pp. 75*ff*.

[3] i.e. the annual conference of representatives from the Panel Committee of doctors from each insurance area, not merely British Medical Association members but all practising health insurance doctors.

mittee in the British Isles which would have passed such a resolution for discussion at the conference, and I am perfectly certain that no conference that I have attended would have endorsed such a proposal. That should be sufficient evidence that any statements made to this effect on your side of the Atlantic are irresponsible statements of individual practitioners, or, which is much more likely, have been culled from the pages of the *Daily Mail* or some other of our less responsible press."

In two instances, the suspicion was voiced in good faith that the suggestion in America that "certification" should be the function of government employees, rested on a project of "work relief" for unemployed doctors!

VIII

DISCIPLINARY PROCEDURE AFFECTING THE HEALTH INSURANCE DOCTOR

THE doctor who is accused of violating his health insurance obligations must, of course, answer the charges, as must an insured person whose conduct is complained of by the doctor. Complaints about medical service or certification are brought by the patient, the insurance society, the Insurance Committee, or the Panel Committee of doctors. The doctor who feels that an insured patient has unreasonably imposed upon him may in turn make formal accusation.

Originally all complaints against an insurance doctor, except the accusation of extravagance in prescribing, were investigated by a subcommittee of the Insurance Committee. Now,[1] however, if investigation by the Ministry indicates a prima facie case of undue laxity in certification, extravagance in prescribing, or failure to keep proper clinical records, the matter is referred to the purely medical Panel Committee, representative of the health insurance doctors in his area. The doctor whose conduct is under question is given a statement of the specific charges and is invited to present his case orally or in writing, as he chooses. If the case involves the adequacy of medical records, a representative of the Minister is entitled to attend the hearing and give evidence.

If the Panel Committee decides that the doctor has been guilty of lax certification or has failed to keep medical records properly, it so notifies the Minister (and the doctor in ques-

[1] If an insured person appeals against the decision of his Approved Society to withhold sickness or disablement benefit, the doctor must attend the appeal and testify. He is paid a fee and allowed liberal travelling expenses. These appeals are infrequent.

tion) and recommends what amount, if any, should be withheld from the practitioner's remuneration. If the Panel Committee decides that there has been extravagance in prescribing,[2] the Insurance Committee, as well as the doctor and the Minister, must be notified.

The standard which the Panel Committee is expected to use in determining whether there has been extravagant prescription is well summed up by the following statement made by the Appeals Board in an "excessive prescribing case":[3]

"In interpreting the term 'reasonably necessary,' allowance must be made for the personal element. It is clear that, if insurance practitioners are to retain the independence of judgment that is necessary for the best treatment of their insured patients, rigorous conformity with the view of the majority of the profession or of a committee of the profession must not be required of them; and so that the progress of medical science may not be checked, scope must be allowed for the exercise of initiative and for suitable experiment. The enforcement of stereotyped methods and standards of prescribing would be prejudicial both to the immediate interests of patients under treatment and to the ultimate interest of insured persons generally.

"There is, however, no inherent connection between the cost of drugs and their therapeutic efficiency. That exercise, therefore, of individual judgment which leads a practitioner in one class of cases to use a drug costing more than that which the majority of the profession would think more suitable, may equally, in other groups of cases, lead him to adopt less expensive forms of treatment than others, and thus, on balance, it will not seriously effect the average. Where particular instances show a proneness to the use of drugs more expensive than most practitioners think suitable

[2] Extravagant prescribing investigations were turned over to the Panel Committee in 1913; certification cases in 1930; and clinical record cases in 1936.

[3] *Inquiries and Appeals under the Medical Benefit Regulations*, Vol. I, part III, case IV, at pp. 50-1.

for treatment of cases of the same kind, or in quantities substantially greater than other practitioners think suitable, and this is accompanied by a very high average cost of prescribing over the practice as a whole, only the strongest evidence can demonstrate that that high cost is due solely to the conscientious exercise of a discriminating judgment, which takes account not only of the patient's needs but also of considerations of economy, so far as consistent with the patient's welfare. In the absence of such proof the conclusion is inevitable that other causes have been at work, such as carelessness or indifference to cost, routine habits of prescribing particular combinations, doubtless appropriate for some cases but applied without discrimination to others, or errors of judgment arising through insufficient appreciation of a practitioner's responsibilities; and that in the result the cost has exceeded what was reasonably necessary for adequate treatment."

The doctor in any one of the three situations investigated by the Panel Committee may appeal to the Minister against the committee's decision, and the Minister then must appoint a person or persons (not more than three, not officers of the Ministry, and at least one a doctor) to hear the appeal. If, moreover, the Minister is not satisfied with the Panel Committee's decision, he may appoint such person or persons to determine the issue. If the case concerns extravagant prescribing, the Appeals Board's decision goes to the Insurance Committee, which in turn makes a report to the Minister.

The Minister then takes under advisement the recommendation of the Panel Committee, the Appeals Board, or the Insurance Committee, as the case may be.

Before deciding upon a penalty, the Minister may and usually does consult with an Advisory Committee. In "overprescribing" and "clinical record" cases, this will be the Medical Advisory Committee, composed of the Chief Medical Officer (or deputy), two other medical officers, and three practitioners who either are or have been insurance doctors. The latter must be selected from a panel[4] offered by the Insurance Acts Committee of the British Medical Association. If a certification matter has been at issue, the Minister may take counsel instead from an Advisory Committee of thirteen (appointed by himself), representative of Approved Societies and Insurance Committees, as well as insurance doctors.

Before deciding to withhold any penalty sum from a practitioner[5] who has not had his case before the Ministry on appeal, the Minister allows him to protest either orally or in writing. If the doctor elects to protest orally, a physician from a list or panel furnished by the Insurance Acts Committee sits as one of the persons who conducts the hearing for the Minister, and both the Insurance Committee and the Panel Committee may have representatives present.

The Minister himself ultimately decides what penalty, if any, shall be inflicted. There is no appeal from this decision.

The body which in the first instance investigates the complaints other than the three types which go to the Panel Committee is the medical service subcommittee of the Insurance Committee of the area.[6] This subcommittee has a chairman and from six to ten members and is composed of an equal number of doctors and insured persons (at least one must be a woman). Normally one of the doctors is appointed by the local Medical Committee[7] and the others by the Panel Committee. If any of these appointments are not made, the Insurance Committee selects a representative doctor for the vacancy.

The chairman must be one of the "neutral" members of the Insurance Committee, someone not immediately in-

[4] This is a committee representative of the panel practitioners.

[5] In reality from the Insurance Committee, which in turn deducts from the next check of the practitioner in question.

[6] See Introduction, p. 5 *supra*.

[7] A committee representative of *all* doctors in the area (often one committee serves as both the Medical and the Panel Committee).

volved in the scheme, i.e. not either a doctor, druggist, insured person, or insurance society officer.

Complaints to be heard by the medical subcommittee include charges of neglect, improper fees, and similar misconduct, as well as illegal certification.[8] Complaint made by one doctor against another involving medical service to insured persons, at the desire of the doctors involved, may be settled by either the Panel or Medical Committee, instead of being referred to the medical service subcommittee. The Insurance Committee may refer the matter to its own subcommittee if it chooses to do so. If the former is done, the doctors must be prepared to abide by the decisions, as such committees have no power to enforce their decisions. Complaints made by the doctor against an insured person go to the medical service subcommittee, and the Insurance Committee may fine an insured person and even suspend his medical benefit for imposition upon his doctor.

Complaints against a doctor must be made ordinarily within six weeks of the episodes upon which they rest. Permission of the Minister of Health is required if more than two months have elapsed, unless the accused person consents to the delayed investigation.

If a complaint involves both a doctor and a pharmacist or is concerned with a pharmaceutical question, a "joint services subcommittee" (composed of two pharmacists, two doctors, and two insured persons) makes the original investigation.

The proceedings of the subcommittees are governed by rules adopted by the local Insurance Committee and approved by the Ministry; these rules must provide for due notice to all persons concerned and for exchange of the statements made by both the accusing and defending parties. Complaints must be specific and detailed. No generalizations

[8] e.g. giving a certificate when he has not seen and examined the patient within twenty-four hours. See p. 49 *supra*. If, however, the Panel Committee is investigating the certification *policy* of a doctor, the Medical Service subcommittee may not act on a certification complaint involving him.

are permitted. The hearing must be private and no paid attorneys are permitted, although the committee may authorize non-professional, unpaid assistance. Usually the witnesses may be present only when actually testifying.

Insurance Committees have funds available for paying travelling and out-of-pocket expenses incurred by the doctor in attending a hearing of the medical service subcommittee, and unless special reasons exist for not allowing such expenses, the doctor may expect to be reimbursed.

The subcommittee's report to the Insurance Committee must include the facts it regards (1) as established and (2) as inferable from those established; it gives (3) in addition, a recommendation of any action it believes should be taken.

If it is recommended that part of the practitioner's remuneration be withheld, the committee is expected to name the suggested penalty sum, if they are all agreed, or, failing this, at least to indicate whether or not the sum withheld should be substantial.

The Insurance Committee must accept the subcommittee's finding of actual facts, but not its conclusions with reference to those facts. For example, a certain subcommittee reported that a doctor had been guilty of neglect. This was a judgment or conclusion with which the larger Insurance Committee was privileged to disagree. The further statements in the subcommittee's reports, however, (to wit, that the insured person called the doctor at nine in the morning, again at noon and twice in the evening, etc., and that the doctor did not visit him in response to these calls until two days and nights had passed), were findings of facts that the Insurance Committee was obliged to accept.

After considering the report of the medical subcommittee on a case involving a charge against a doctor of neglect of a patient, the Insurance Committee may in its discretion proceed to penalize the doctor in several ways. It may require him to pay, usually by deduction from his quarterly check, any expenses[9] other than those of the subcommittee, to which the Insurance Committee or the insured person or his family, have been put because of the doctor's neglect. For example, if the insured person, because of neglect by his panel doctor, has had to call another physician, the committee may require the insurance doctor to pay such physician's fees.

Or the committee may, after consultation with the Panel Committee of doctors, require the doctor to reduce his insured list below the usual limit, on the ground that he has demonstrated that he is not capable of giving adequate treatment to so large a number.

It may report to the Ministry of Health that the doctor has not fulfilled his insurance obligations and is not entitled to full remuneration.

Finally, and this of course would be done only in the face of the gravest misconduct, it may report to the Minister that the doctor should be denied the privilege of health insurance practice.[10]

The doctor is allowed to appeal to the Minister from a decision of the Insurance Committee in any case except that which terminates with a recommendation that his name be stricken from the Medical List, i.e. that he be denied the right to insurance practice. Appeal from the latter recommendation is automatic: the Minister is obliged to hold a formal judicial inquiry.

In the case of these appeals, the Minister may assess costs against either party. He may dismiss an appeal if he decides after considering the application that no reasonable grounds for appeal exist. He may, moreover, decide the appeal without a formal hearing. As a rule, however, he appoints one, two, or three persons to hear the appeal—which is usually conducted in the locality of the dispute—and to present a report. If the doctor has been found guilty of "negligence,"

[9] Medical Benefit Regulations 35 (2) (b). (*Stat. Rules and Orders*, 1936, p. 1903.)

[10] Literally, "that the continuance of the doctor on the medical list is, in their opinion, prejudicial to the efficiency of the service."

one of the Appeals Board must be a doctor selected, as in cases investigated by the Panel Committee, from the panel[11] list presented by the Insurance Acts Committee of the British Medical Association. "Negligence" is defined as "Failure to exercise reasonable skill and care in the treatment of a patient; failure to visit or treat a patient when necessary; failure to order or supply any necessary medicine or appliances for the use of the patient; failure to advise the patient as to the steps which should be taken to obtain treatment which is not within the scope of the Terms of Service."

Appeal proceedings are judicial proceedings similar to British arbitration proceedings, i.e. witnesses may be subpoenaed, parties examined on oath, and points of law may be referred to the court for decision. Lawyers are permitted on both sides. The Appeals Board at its discretion may rehear all the facts or just those at specific issue. Contentions and facts not presented at the original hearing may not be presented at the appeal without special permission of the appeal body; this permission is seldom granted. The decision of the Minister as to guilt or innocence is made after he has considered the report of the Appeals Board and is final.[12]

As in cases originating with the Panel Committee, if the practitioner condemned by the medical subcommittee has not appealed, he is allowed a hearing by the Ministry before penalization by the latter. If an accusation of "negligence" has been at issue, the Medical Advisory Committee[13] must be consulted before action is taken. In other disputes involving a doctor, consultation with this Advisory Committee is optional and, as in matters investigated by the Panel Committee, as a matter of practice is usual. Such cases are when the doctor has: (1) charged fees to insured persons; (2) issued

[11] See note 4 *supra.* i.e. the panel submitted by the Insurance Acts Committee from which members of the Medical Advisory Committee are selected.

[12] The Panel Committee is not entitled to be represented at the hearings of appeals. The practice is, however, for appeal tribunals to permit such representation.

[13] This was added by the Regulations of 1927. For membership of Advisory Committee, see pp. 64-5 *supra.*

certificates irregularly; (3) failed to fulfill his obligations to the Regional Medical Staff.

Determination of Penalties

The usual penalty is a "surcharge," i.e. a deduction from the doctor's remuneration. Literally, the Ministry deducts this from the sum allotted to the Insurance Committee of the area in which the penalized doctor practises. The committee in turn deducts it from the next quarterly check of the doctor in question.

The amount of penalty is related, of course, to the additional cost which the doctor's misconduct imposes upon the funds of the insurance scheme. It is not, however, rigidly adjusted to the figure which the Panel Committee judges to be the measure of such cost. In fact, it is usually considerably less than this figure. To quote from a pertinent decision of the Appeals Board:[14] "Provision for deduction is largely intended to act as a deterrent; extravagance while its existence may be indisputable is nevertheless very difficult to bring to a precise measurement in pounds, shillings, and pence." And again: "Such deductions are to be regarded as primarily deterrent in their object, and for this purpose considerations must be taken into account which cannot be satisfactorily evaluated by any purely arithmetical process, e.g. considerations as to the good faith of the doctor in actions which perhaps were mistaken."[15]

Exclusion from Insurance Practice

The suggestion that a practitioner be denied the right to insurance practice, which, as previously stated, may come from an Insurance Committee after a subcommittee investigation and report, may also emanate from either of two

[14] *Inquiries and Appeals under Medical Benefit Regulations*, Vol. I, part III, case IV, cmd. 8549 (1917).

[15] Harris and Sack, *Medical Insurance Practice*, p. 101.

medical bodies, the local Medical Committee or the local Panel Committee in an insurance area.

On receiving such a recommendation, the Minister must, and in certain situations[16] *may*, set up a Committee of Enquiry, consisting of an attorney and two practising physicians. This committee holds a formal judicial hearing, with witnesses on oath and (usually) legal counsel on both sides. The resulting report in turn must be referred to the Advisory Committee of doctors for consideration and recommendation before the Minister can take action. Further, the doctor is permitted to present to the Minister written evidence as to his character and his professional reputation.

A doctor who is thus removed by the Minister from his local panel or medical list may not again engage in health insurance practice or assist a health insurance practitioner anywhere in Britain without first getting permission from the Minister. The Minister's decision is final and there is no right of appeal to the courts.

Frequently, the Minister determines upon lesser penalties, instead of complete removal from the medical list. He may order temporary suspension and probation, or, the usual penalty for most offenses committed, the withholding of part of the doctor's normal remuneration. The customary penalty sums withheld range from $5 to $500, although occasionally the amounts are smaller or larger.

In summary, it may be said that the significant fact about the disciplinary control of the health insurance doctor is that representative medical opinion plays a most important rôle in all its procedure. Thus complaints are investigated in the first instance by either a purely medical or a semi-medical body (the Panel Committee or the medical subcommittee of the Insurance Committee); in appeal proceedings on Panel Committee cases or cases involving negligence a medical representative is a member of the board; and the Ministry of

[16] When such recommendation comes from some other source or when the doctor is convicted of crime.

Health may sit in judgment upon the doctor's performance of his medical duties only after consultation with still another representative medical committee (the Medical Advisory Committee). Further, when the right of the doctor to health insurance practice is at issue, penalization may come only after formal trial by a predominantly medical board (three members, two of whom are physicians) supplemented by consultation with an Advisory Council of doctors (i.e. the Medical Advisory Committee).

All the medical men involved in administration of the disciplinary procedure are either selected directly by practising physicians or by their representative committees, or are chosen by administrators from lists drawn up by the British Medical Association's Insurance Acts Committee. Over nine-tenths of the latter committee are representatives elected by the health insurance doctors; the remaining members are officers of the British Medical Association.

The complaints registered against health insurance doctors have been very few in number in rural areas. Even in London and the other large cities where most of the abuses have occurred, there have been surprisingly few formal accusations. This seems to be caused not exclusively by the native wickedness of a few of the urban practitioners, but in part by the fact that the insured person in some of the villages and the rural areas finds it harder to bring charges against the doctor. Often the latter is the only man available. Still more often, the insured worker lives in such close proximity that he cannot get away from his doctor. He is, therefore, naturally reluctant to enter formal complaint against him.

The County of Buckingham, with an insured population of 110,000, has had only sixty medical service cases in the last eighteen years, and most of them in the earlier part of the period. Derbyshire, with a quarter of a million insured, has averaged about two complaints a year over the entire twenty-five year period of the act's functioning. The City of Manchester, with an insured group of from 320,000 to

350,000, has totalled 127 formal complaints by patients in the twenty-five years of health insurance.[17] Frequently the same doctors were involved year after year.

In 1934 eighty-one, and in 1935 a total of one hundred and one, English health insurance doctors were penalized by having part of their remuneration withheld. The penalties totalled £626 6s. ($3,131.25) in 1934 and £779 15s. ($3,898.75) in 1935. The misconduct established was as follows:

Specific Misconduct	*Number of Cases* *1934*	*1935*
Negligence[18] in medical service	14	6
Failure to keep proper medical records	20	39
Failure to give information to the R.M.O.	16	18
Violation of medical certification rules	15	16
Failure to return medical records to Insurance Committees	6	13
Improper charging of fees to insured persons	4	2
Combination of violation of medical certification rules and improper charging of fees to insured persons	1	—
Failure to arrange for treatment of patients during absence from practice	1	—
Leaving practice without giving notice to Insurance Committee of withdrawal	1	—
Issuing prescriptions to persons in dispensing list	—	1
Refusing R.M.O. access to medical records	—	1
Excessive prescribing	3	5*

*Plus 1 pending.

The negligence cases, which are in one sense the most serious, as compared with the record in 1932 and 1934 are given below:

1932	22
1933	8
1934	14
1935	6

[17] Fifteen in the years 1931-1935 involving, however, only nine doctors; six in 1935, involving, however, only two doctors. Information furnished by Mr. Dewhurst, assistant secretary of the Manchester Insurance Committee, July 1936.

[18] See first paragraph, p. 69 *supra*, for conduct constituting negligence.

The largest single amount withheld from one doctor was $300 in 1934 and $500 in 1935; $250 was the penalty in another 1934 case; $125 was deducted in several instances in both years. The great majority of penalties, however, in both 1934 and 1935 were in amounts of less than $25.[19]

Between 1913 and March 31, 1936, there were 99 formal charges involving the removal of doctors from the panels, or, as it is termed in the British reports, "Representations that the continuance of doctors on the medical list would be prejudicial to the efficiency of the insurance medical service."[20]

[19] Ministry of Health, 16th *Annual Report*, pp. 255-6; 17th *Annual Report*, pp. 196-8.

[20] In the case of the first 30 Representations, 20 of the Inquiries resulted in removal of the doctor from the panel (in one case the Insurance Committee was authorized to restore the doctor in a year). In one case the doctor was permitted to resign; in three cases he was fined; in one case he was found innocent and in the remaining 5 cases no official action was taken.

The removals were for such offences as (1) absence without arranging for a deputy; (2) dispensing drugs under an assumed name; (3) charging for medicines; (4) intoxication during office and visiting hours; (5) signing certificates without examination; (6) neglect of patients; (7) improper financial connection with a druggist.

IX

ATTITUDE OF THE BRITISH MEDICAL ASSOCIATION AND PANEL DOCTORS

The Part Played by the British Medical Association

IN keeping with the democratic principles of British political and industrial life, an agency of the practising medical profession for "collective negotiation" actively participates in the Health Insurance Administration. The agency which is accepted by the Minister of Health as representative of the health insurance doctors is the already mentioned Insurance Acts Committee of the British Medical Association. All rulings which affect the remuneration of the panel physicians or any of their obligations under the health insurance act are made after consultation with this committee. The Ministry is pledged, moreover, to this procedure as a matter of permanent policy.

The Insurance Acts Committee is a central committee composed of representatives of all parts of the country. The representative member in each area keeps in constant touch with the local Panel Committee and with the local Medical Committee when there are two separate committees. By frequent meetings in London, the central committee members form a clearing house of information about conditions of health insurance practice throughout the country.

In addition, the Insurance Acts Committee calls an annual conference of representatives of the Panel and Medical Committees of all insurance areas. These representatives need not be members of the Medical Association. All health insurance problems which are of interest to the profession are debated at these conferences. All suggestions forwarded by the local Committees (Panel and Medical) are discussed and resolutions are passed upon all subjects affecting the medical profession in its relation to health insurance. These resolutions are accepted as binding upon the Insurance Acts Committee. Thus the committee makes itself in as real a sense as possible the mouthpiece of the great body of insurance practitioners.

The participation of the medical profession in the disciplinary procedure affecting the health insurance doctor has been outlined. It remains to repeat and to stress the fact that the Insurance Acts Committee appoints the panel of doctors from which selections must be made to the boards which hear appeals and to the Advisory Committee that figures in the final disciplinary procedure of the Ministry previously described. From this panel are chosen also the doctors who sit with Ministry officials in hearing the oral protests of a doctor before penalization by the Minister. The Insurance Acts Committee thus has the ultimate responsibility of choosing the medical men who guard the interests of the profession in the final stages of the penal machinery of the health insurance scheme.

The Insurance Acts Committee is, however, much more than a negotiating and protective agency for the individual practitioners. It is dedicated to the policy of cooperation with the government officials in making the social insurance system as fine and effective as possible. It furnishes efficient machinery for detection and correction of friction points and other dangers inherent in the general system. Indeed, too much could scarcely be said for the part which the organized medical profession has played, especially in the last twelve to fifteen years, in the evolution of the British social insurance scheme into the smooth system it has become.

It would have been easy for the Insurance Acts Committee to take a narrow view of its obligations. Instead, it has adopted as a fundamental principle the doctrine that the interests of the profession can be well served only by making the insurance scheme achieve the social objectives which prompted the enactment of social insurance legislation. To that end the Insurance Acts Committee has worked unceasingly to educate the profession to an appreciation of (1) the importance of a scheme which provides the bulk of the working population with security in time of illness; (2) the soundness of the democratic principle which opened health insurance practice on equal terms to the whole profession; (3) the solemn obligation which the profession, in return for this privilege, is called upon to shoulder by performing with scrupulous care the key services to the scheme upon which its functioning rests; (4) the need of sympathetic understanding of the difficulties confronting the Insurance Committees as well as those of the Approved Societies which dispense the cash benefits; and, finally, (5) the fact that the dignity and well being of the whole medical profession are involved in the task of keeping health insurance practice on the highest possible professional level.

Differences in the standard of health insurance practice and other practice, separate offices and "side doors" for insurance patients, and other discriminatory practices of the early bickering years have been stamped out, largely through the efforts of the Insurance Acts Committee.[1] In similar fashion, relations between the Approved Societies and the profession have so changed that the armed neutrality and mutual distrust which characterized the early period have been replaced by friendly cooperation and mutual respect. The committee has been working hard to aid the administrative agencies in their drive against the excessively high cost of drug benefit, which has been the chief administrative concern of the last few years. In every region representatives of the committee stand solidly back of the Ministry and the network of administrative officials in their efforts to eliminate waste and improve the insurance services. It is also looking toward expansion of the scheme in the near future

[1] A lion's share of the credit due the Insurance Acts Committee is attributed by practitioners in all parts of England to Dr. Guy Dain, chairman of the Insurance Acts Committee for eleven years (1924 to November 1935).

and is recommending particularly the extension of services to the families of the insured workers. It is urging medical schools to include broad courses on health insurance in the required curriculum, so that doctors may practise with an appreciation of the social and economic background of the health insurance scheme.

The committee has successfully sponsored in several areas the organization of the free "family doctor" medical services which are publicly provided for the destitute along lines similar to that in effect for the insured population. Free choice of doctor is permitted and the practitioners are paid on a capitation basis, thus giving the dependent poor the same relation to their family doctors that is enjoyed by the insured workers.

Why the Panel Doctors Approve of Their Health Insurance Arrangements

The change in attitude of the medical profession from bitter hostility when legislation was being discussed in 1909 and 1910 to that of enthusiastic approval after years of actual experience with health insurance is attributed by practitioners to a variety of causes. Dominant among such causes they concede to be the great economic improvement which health insurance brought to doctors. Few human beings are indifferent to such improvement and it is not surprising that medical prejudice against legislation which resulted in greatly increasing medical incomes should tend to disappear. Panel doctors in city, town, and shire emphasized this increase in medical income, estimating that income from practice among wage-earning groups had doubled and even trebled as a result of health insurance.[2]

[2] There is a mistaken assumption widely entertained in the United States that the practice included in the health insurance scheme had all been on a contract or "club" practice basis before the scheme was adopted. In fact, as the B.M.A. protested in 1912, club practice had not even reached persons of more than £2 a week earning capacity, while the act included all manual workers and non manuals up to £3 a week. See the 1912-1913 *Report of the National Insurance Commissioners*, 1913, cmd. 6907, Part I, pp. 124-57.

Almost as important as the betterment in amount of income is the *regularity* of income which the insurance arrangements have brought the profession. The fact was stressed repeatedly that doctors appreciate security as much as anyone. The steady quarterly check from the insurance committee, in bad employment periods as in good (for health insurance contribution arrears are excused when caused by unemployment), in intervals of good health as in time of epidemic, serves as an income backlog to the practitioners. This stability apparently is valued by the doctors whose practice is only partly among workers' families as well as by those who draw all their patients from wage-earning groups. Freedom from the nuisance and bother as well as from the uncertainties and losses of bill collecting, was universally welcome to town and country practitioners.

It would be unjustified, however, to lay undue emphasis upon the economic factor. The panel doctor with real professional feeling (and this means more of the panel than not) approves of health insurance because he has found that it greatly improves his relations with his patients. Health insurance, he has discovered, removes a barrier which in "private practice," among low-incomed groups, stands in the way of full realization of the fine relationship that "doctor and patient" should achieve. Undeterred by the menace of a bill which he could not afford as an individual, the insured patient can see his doctor when he needs to. Similarly the panel doctor, without concern as to how much the patient can pay, and without courting suspicion that he is "running up bills," can supervise his patient as he deems best.

As one practitioner put it: "You have no idea how pleasant it is for me to be able to drop in as often as I like on my poor patients who are chronics. In the old days I couldn't possibly have done it. Either I'd have had to charge them for calls, which wouldn't have been honest, or else I'd have had them feeling no end uncomfortable and grateful to me for not charging them. Now we meet on the finest level. I'm their doctor and paid for and that settles that. So I can step in for a minute or two every day if I like and it does us both a lot of good. It's a grand profession, medicine, if you give it a chance." And another: "Health insurance has transferred to the authorities the 'business' part of private practice and left us free to practise our profession, which, after all, is medicine, not bill-collecting."

Capitation is believed by the practitioners generally to be the only satisfactory basis of health insurance remuneration. This is partly because they have found it to be administratively satisfactory, in contrast to the fee system's problems, typified by the Manchester experience. At the same time, capitation preserves what British doctors deem to be the desirable elements of the private practice fee basis relationship. These elements they list as selection of doctor by the patient and patient by doctor, the right to sever this relationship if it proves unsatisfactory, and variation of remuneration of the doctor in accordance with his ability to attract and satisfy patients. A "salary" basis would, of course, be attractive from the standpoint of certain purely administrative needs, for the authorities could control the type of service rendered by dismissal of the doctors who did not maintain adequate standards. Such a basis, however, doctors find basically objectionable for two main reasons. It would restrict health insurance practice to selected appointees rather than opening it to the profession, and would, general medical opinion is convinced, tend to lower the average standard of professional service by removing the economic stimulus which "keeps the doctor on his toes."

Capitation, in brief, while obviating on the one hand the difficulties which the Manchester and Salford doctors encountered to their sorrow,[3] has the advantageous element of the fee system of penalizing the doctors who neglect their patients, for the capitation sums move with the patients away from such medical men to more conscientious practitioners.

[3] See pp. 41-5 *supra*.

The doctors find further virtue in capitation, virtue which is of importance to the community as well as to the individual doctor and his patients. They maintain that capitation promotes "good medicine" because it is a system of remuneration based on *responsibility for the care needed* by patients rather than on items of service rendered to patients. It is therefore to the doctor's interest to keep his patients well. Under the fee system, the practitioner's financial interests lie wholly in the reverse direction. With capitation, if a patient gets seriously ill, the doctor as well as the patient suffers for it. Scrupulous attention to the minor things which lead to the serious is as important and time-saving to the doctor as to his patient, while in no way reducing the doctor's income. In short, the doctors have found in capitation a direct stimulus to preventive medicine.[4]

A homely illustration of this stimulus is embodied in the little printed card which one of the leading panel doctors in Buckinghamshire gives each person who joins his medical list. It reads, "If you get a cold, come to see me. Let *me* decide whether it is important. I should rather treat a cold than pneumonia."

Another improvement brought by health insurance and emphasized by doctors is the elimination of the lowest element of the profession. This, doctors of long experience explain, is not caused exclusively by their elimination through the disciplinary control previously described, although that process undoubtedly has removed a substantial number of the grossly unfit practitioners. It is attributed also to the successful competition of better men who, through health insurance, now offer their services to the lowest paid workers. Before the institution of panel practice, in the days when doctors had to collect for their services directly from their poorest patients, apparently very few medical men of stand-

[4] See emphasis of this stimulus to preventive medicine entailed by health insurance practice, in the *Annual Reports* of Chief Medical Officer of the Ministry of Health, 1919-1925.

ing would endure such conditions as were indicated by "Eastside" London practice.[5] The 1914 report of the National Health Insurance Commission refers to the fact that some of the poorest boroughs actually had no resident doctors at all before the health insurance law and that all of the poor districts had been most inadequately served. To put it in the words of a distinguished London practitioner of more than thirty years' personal experience with such conditions:

"In the old days, well trained, fine young doctors could not stomach practice in these areas. All too often it meant the hardest kind of work and either no fees or a shilling or two, raised sometimes by the patient's pawning the blankets from his bed. Health insurance changed all that. Now the worst type of doctor, who was the usual Eastside practitioner, is a thing of the past. In some cases he has been found guilty of gross abuse and has been denied the right to insurance practice. For the most part, however, he has been competed out of existence. Good medical men are now in the field and poor people appreciate decent service as much as anyone else and will take it when they can get it."

Search for a health insurance practitioner who would be willing to exchange his insurance arrangements for those of American "private practice" was unrewarding. There may be such a person[6] but diligent investigation in towns and shires failed to reveal him. Men can be found everywhere, although surprisingly few on the whole, who will complain, or "grouse," as they term it, about this or that detail of insurance practice, but all its basic principles are universally approved by the men who speak from experience. Their only real objection is that there is too little health insurance—i.e. that not enough of the available health services have been organized on the insurance basis, and not enough people are included in this part of the social insurance program.

[5] The very poor never joined "medical clubs."

[6] Although several well known practitioners were willing to lay substantial wagers that "no such foolhardy soul would be found in the whole of Britain."

Extensions of Health Insurance Recommended by the British Medical Association

The organized medical profession is, as stated, sponsoring amendment of the Health Insurance Act to provide medical benefit for the dependent families of the insured workers. For this proposal there is strong lay support. Indeed, agitation for such a change has been widespread for many years. It has long been quite widely conceded that lack of medical service for workers' families is one of the great shortcomings of the British scheme, placing the latter in uncomfortable contrast with the systems on the Continent. The Royal Commission of 1924, however, did not consider it even a proper health insurance benefit, stating "Medical opinion for dependents should form an integral part of any scheme of general health service, administered by the local authorities."[7] A recent government document adds a degree of current official sanction to the urging of the medical profession for "family benefits" in the health insurance scheme. A Scottish Departmental Committee concluded in the summer of 1936 an inquiry[8] into the functioning health services, with a strong recommendation that such benefits be extended in the immediate future.

There is a fairly strong tradition that things which are under consideration for the United Kingdom as a whole are "tried out" in Scotland. General medical and lay opinion has it that the action of this Scottish body is prophetic of what may be expected in the course of time from the Ministry at Whitehall.

In 1936, however, more official interest seemed to be given to the other proposed extension, i.e. the broadening of medical benefit itself to include specialists' services. The lack of specialists' service and laboratory analysis service has been governmentally recognized as a major limitation of

[7] Report of the Royal Com. on National Health Insurance, 1926, cmd. 2596, p. 163.

[8] Published as the *Report of the Committee on Scottish Health Services*, Department of Health for Scotland, 1936, cmd. 5204.

medical benefit practically since the institution of the scheme. In 1914 provision for beginning such services was actually included in the budget, and moneys were voted by Parliament.[9] This project, however, was lost in the World War activities and continued to be lost in the post-war adjustment difficulties. After the organization of the Ministry of Health in 1919, the Chief Medical Officer's annual reports saw to it by repeated emphatic reference that the question at least was not forgotten. The Royal Commission on National Health Insurance of 1924, moreover, placed specialist, consultant, and laboratory services at the head of the list of immediate statutory additions required for the health insurance scheme.[10] This recommendation, however, was not put into effect.

The organized medical profession now is actively urging the "modernizing" of medical benefit. This proposal involves the participation of a group of physicians who are not now in health insurance practice, rather than, as in the case of the "family benefits" proposal, an extension of the group served on an insurance basis by the panel doctors now practising. Difference of opinion about the details of this extension concededly exist, but the organized profession, at least, is beginning to sponsor concrete proposals.

A special committee of the British Medical Association in the spring of 1936 worked over the questions involved in initiating the broadening of medical benefit beyond family doctor service. In June 1936, it presented a report in which its general conclusions were summarized in the following resolutions:

> 1. That consultant, specialist, and pathological services should be available as an integral part of medical benefit under the National Health Insurance Acts—that is, for all those insured persons entitled to medical benefit.

[9] Great Britain Royal Com. on National Health Insurance, 1926, cmd. 2596, p. 123.

[10] *ibid.*, p. 173.

2. That such services should be available only on the recommendation of the general practitioner in attendance.

3. That there should be free choice of consultant, specialist, and pathologist by the general practitioner in attendance.

4. That in the provision of pathological facilities use should be made of the services of pathologists working in private laboratories and of pathologists working in the laboratories of voluntary and council hospitals, the pathologist receiving adequate remuneration for his work.

5. That in areas where a pathological service is provided for the community by the local authority the use of this service by insured persons should continue, the pathological services under this scheme supplementing and augmenting the local authority facilities so as to make the pathological services available to insured persons complete.

6. That the consultant and specialist services in medicine, surgery, gynecology, and the special branches should be such examination and advice as can be given at a *single consultation*, together with a report, where necessary, for the information of the attending practitioner.

7. That the inclusion of the name of a medical practitioner in the list of available consultants, specialists, and pathologists should be subject to satisfaction of one or more of the following criteria: (a) that he has held hospital or other appointment affording special opportunities for acquiring special skill and experience of the kind required for the performance of the service rendered, and has had actual recent practice in performing the service rendered or services of a similar character; or (b) that he has had special academic or post-graduate study of a subject which comprises the service rendered, and has had actual recent practice as aforesaid; or (c) that he is generally recognized by other practitioners in the area as having special proficiency and experience in a subject which comprises the service rendered.

8. That the decision in regard to the eligibility of practitioners for inclusion in and continuance on the list of available consultants should rest with a properly constituted professional body.

9. That the list of available consultants, specialists, and pathologists should be divided into two sections, the first consisting of the names of those who are not engaged in general practice in any form but practise exclusively as consultants,

specialists, or pathologists, the second consisting of the names of those not so practising.

10. That the remuneration for the work should be on a per item of service basis.

11. That the organization of such services should be on lines strictly comparable to those obtaining in private practice to-day, the consultant receiving the patient in his own consulting rooms or attending at the patient's home, as the case may be.[11]

If the essence of these conclusions were adopted, obviously a mere first step would be taken in the direction of making "specialist service" a normal part of the insurance benefits. This first step, moreover, does not propose a step into completely untried territory. The British Medical Association has been sponsoring such a scheme on a voluntary basis.[12] It circulates among the members a List of Consultants in London who are willing for a greatly reduced fee (one guinea, $5.25) to see insured persons and others in proved similar economic circumstances, and to give "such examination and advice as can be given at a single consultation, together with a report, when necessary, for the information of the private practitioner. The names included in the list are arranged under the following headings: medicine; surgery, obstetrics and gynecology; ophthalmology; dermatology; oto-rhino-laryngology; pathology, radiology; physical medicine."

The organized profession is well aware that in this age of specialization in medicine, lack of specialist services in the health insurance set-up is a lack so serious as to promote development of alternative public response (chiefly through county action) to the need of the workers for more than mere "family doctor" services.

It is equally aware that the failure of health insurance to provide family benefits also is stimulating the growth of

[11] Published in "Medical Insurance Service Week by Week," supplement to the *British Medical Journal*, June 27, 1936, p. 358.

[12] Just as, pending addition of the "family medical benefits," it has sponsored voluntary insurance at very low cost, for the dependents of insured workers. See p. 41 *supra*.

alternative public services to meet the medical needs of the workers' families. The risk of unacceptable arrangements from the medical point of view, which "alternate" public action (i.e. County Council Medical Services extensions) might bring, is a recognized risk in medical circles. Content with the medical service pattern worked out in the health insurance scheme, the British Medical Association is doing what it can to extend this pattern.[13]

[13] Formal proposals were made by the British Medical Association for coverage extensions and amplification of health insurance services, in its 1930 *Proposals for a General Medical Service for the Nation.* See Supplement, *British Medical Journal,* 1930, pp. 155-82.

X

HOSPITALS AND THE HEALTH INSURANCE PROGRAM

THE omission of hospital benefits not only in the present British health insurance scheme, but even in the plans for its immediate future expansion, is completely at variance with the usual social insurance pattern.[1] It is accounted for chiefly by the so-called "voluntary hospital" tradition, which, though badly shaken and radically altered by post-war developments, still has a strong sentimental place in English life.

The British "voluntary" hospitals started as private institutions existing merely to serve the needs of the sick. In pre-war days, most of these hospitals made no charge whatever for the care of patients. They were, moreover, open to any sick person who chose to go to them, although the well-to-do usually entered "nursing homes," i.e. small private institutions operated by surgeons on a profit basis.[2]

The voluntary hospitals[3] were supported by public contributions, coming not only in great sums from the very rich but also in pennies from the poor. Many a knighthood was granted on the basis of substantial public service through

[1] Great Britain is the only one of twenty-two countries having compulsory systems in 1936 that provided no hospital care.

[2] The British do not term such an institution a hospital.

[3] There was no real "system" of voluntary hospitals. The institutions developed independently of each other with great variety in details. The description here given is that of the usual pattern. Information was gathered in 1932 by the Central Bureau of Hospital Information from approximately three-fourths (74 per cent) of the voluntary hospitals (containing 73 per cent of all the beds in these areas) located outside the large cities. Even at this date less than 5 per cent of the beds were "pay" beds and more than two-thirds of these were in hospitals having under one hundred beds. About one-sixth of the hospitals had had a few paying beds since the inception of the hospital. One-third of the hospitals still had no "paying" beds at all.

endowment of a new hospital wing; special collection days, fairs, bazaars, and many other devices were resorted to for reaching members of all economic groups. It was a treasured tradition that the public, rich and poor alike, each according to his purse, should voluntarily support the hospital, and that the hospital should serve all sick without question of payment. The staff was a "closed" staff, i.e. only its members could attend medical and surgical patients, and as a rule it comprised distinguished physicians and surgeons who gave their services gratuitously or for a nominal honorarium.

Such in general was still the situation when the health insurance law was enacted in 1911. Hospital care was not then on the health insurance dossier. Such care was the business of the great private charitable institutions—the voluntary hospitals.

After the World War, partly because of an aggravation of economic and social changes which had their beginning much earlier and partly because of the long depression which settled down upon Britain, the voluntary hospitals fell into serious financial difficulties. There was first of all great increase in the demands made upon hospitals. This was caused partly by more common demand for the specialist services which could be found in the hospitals and partly by the fact, to use the words of a distinguished British physician, Sir Henry Brackenbury,[4] that "domestic difficulties of managing a long and serious illness at home increased" and "consequently many persons who formerly would never have thought of entering hospitals began to do so. Adequate public contributions and endowments to meet this situation simply were not forthcoming. Despite tremendous effort on the part of the hospitals to keep going on the old basis some change became inevitable."

To quote from the report in 1921 of Lord Cave's Committee on the Voluntary Hospitals: "The practice of asking

[4] Sir Henry Brackenbury, *Patient and Doctor*, London: Hodder and Stoughton, 1935, p. 204.

patients treated at the hospital to contribute to the hospital fund is comparatively new and is a product of the financial stringency. . . .

"Some hospitals make a fixed charge (except in the case of poor people) as a condition of admission, the charge for a patient being generally about 20s. a week. . . . Other hospitals adopt a practice which they consider to be more in accordance with the voluntary principle and after the admission of an in-patient make inquiries as to his means and apply to him (generally through a lady almoner) to make a voluntary contribution to the hospital funds according to his ability. In either case the amount asked for is far below the actual costs incurred."[5]

A Central Bureau of Hospital Information was organized in the early 'twenties for mutual advice and counsel. An "almoner" system was installed by a growing number of institutions to obtain part of their maintenance directly from such of the patients as could afford to pay for care. This system, like all other methods of collecting from patients, of course, struck at the very heart of the voluntary hospital tradition and was instituted only as a last resort alternative to ceasing to function at all.[6]

The almoner system, moreover, never was deemed really acceptable. The compulsory insurance services had educated the British public to the advantage of the insurance technique which collects in advance while people are well and

[5] *Report of Voluntary Hospitals Committee* ("Lord Cave's Committee"), 1921, cmd. 1335, pp. 26, 27. The "almoner" is the equivalent of the American "medical social worker," attached to certain of our public and our private institutions. Her "eligibility" duties require her to investigate the financial circumstances of the prospective patients and advise the hospital management what should be charged for care in specific cases.

[6] The difficulties of handling illness cases in the home—difficulties brought by post-war "domestic help" problems—not only increased the demand on the voluntary hospitals but also stimulated the development of a type of institution new to British experience, commonly spoken of as the "cottage hospital." This is like the ordinary American commercial hospital in that it allows a doctor to follow his patient into the hospital. It is intended as an accommodation for the general practitioner whose patient cannot be cared for at home conveniently. Like other English hospitals, it is not a commercial institution. Charges are, at maximum, to cover costs only.

employs the funds for their care and maintenance when they are ill and unable to work. This was made very clear by Lord Cave's committee report cited above. It outlined at great length several contributory schemes which had just been started, whereby "members" for a small weekly sum were entitled to hospital care without subjection to the almoner (i.e. without further contribution according to their means). And the report concluded: "We have dwelt at length on these schemes of mass contribution, because we are inclined to believe that in the adoption of one or other of them may be found the key to the problem which the voluntary hospitals have to solve."[7]

A proposal to expand the health insurance contributions, however, and to add hospital benefits, obviously needed to round out medical benefit, was conspicuously lacking. Instead, the report stated that "such a proposal would effect a radical change in the position of the voluntary hospitals and require careful consideration."

In like fashion, the later report of the Royal Commission on National Health Insurance failed to recommend making institutional care a statutory benefit, chiefly on the ground that this would be prejudicial to the voluntary hospital tradition.[8]

Ergo came the Hospital Savings movement. Membership in such associations was usually open only to persons in circumstances comparable to those of the wage earners included in the compulsory health insurance scheme. A small sum, as a rule 3d. (6 cents) a week, gives a membership which entitles the family as well as the insured person to benefits. The voluntary hospitals agree to waive "almoner" investigation in cases of members of Hospital Savings Associations, accepting instead from the association a weekly sum of 28s.

[7] *Report of Voluntary Hospitals' Committee*, p. 26.

[8] *Report of Royal Commission on National Health Insurance*, 1926, cmd. 2596, p. 125, and also see evidence of Ministry of Health's representative arguing these same premises. Minutes of Evidence, pp. 105-42.

($7)[9] for each member hospitalized, supplemented by an annual lump-sum grant. According to the information filed with the Hospital Association, the flat-rate payment, small as it is, exceeds the average extracted under the "almoner" system from persons of like economic circumstances. The care guaranteed by the insurance arrangement is limited to ten weeks in a minority of the voluntary hospitals, but is given for an unlimited period as needed in most of them. The Hospital Savings Association of London estimated that the membership in hospital associations in 1936 was about 1,750,000 (dependents bring the benefited group to about 4,000,000), the majority being resident in London and its environs.

With the break in the voluntary hospitals' tradition of gratuitous service and the great increase in the demands on the hospitals has come, quite justifiably, rebellion by many of the staff specialists against the giving of their services without remuneration. The British Medical Association has taken the stand that hospitals that receive funds through contributory schemes should turn over a percentage of the funds to the medical staff. Although the majority of the staffs probably still oppose any break with the long-established custom, it is significant that the younger men tend to approve the stand of the British Medical Association.[10] The tradition of "honorary" service, while dying hard, undoubtedly is dying with the changes in the system that bred it.

Payment of staff physicians is becoming even more immediately inevitable by virtue of current developments in the public hospital field. This recent public hospital development is far more significant in most respects than the changes in the voluntary hospital set-up. As a result of the 1929 amendment of the County Government Act, certain county

[9] The usual supplementary lump-sum "grant" brings it up to 32s. ($8) per week.

[10] See Memorandum 16 issued by the Central Bureau of Hospital Information in 1929 (reprinted in 1933).

authorities have been converting their old "poor law" free hospitals for indigent persons into public "pay" hospitals for the general public. The London County Council has been remodelling, building, equipping, and organizing a splendid health service, which offers to the general public both hospital care and also, through out-patient departments of the hospitals, a great variety of specialist services to persons who are not institutionalized patients. Staffs are paid staffs. Members are partly full-time men but, in deference to pressure from the British Medical Association,[11] are predominantly part-time consultants. Maximum charges to patients are cost charges only (£4 a week, $20, is the full charge), and the weekly charge includes everything—maintenance, nursing, laboratory service, as well as medical and surgical care. "Almoner" service determines the charges for patients who feel unable to meet the normal "full costs." Arrangements with the Hospital Savings Associations similar to those made by the voluntary hospitals (but not so generous)[12] admit members of these associations without cost.

The health service of the London County Council obviously is driving toward the adequate medical servicing of society with an entirely different technique from that of the National Health Insurance scheme. The conspicuous gaps in that scheme caused by the omission of institutional and specialist care in the medical benefit undoubtedly played a part in creating the public pressure leading to this public health service development.

So far, the specialist services offered in the out-patient departments of most of the London County Council hospitals have been restricted largely to previous or contemplated

[11] The organized medical profession waged a bitter battle to secure for all medical men the privilege of following patients into the newly created public hospitals. This privilege was refused on the score that the responsibility of the London County Council in offering institutional care to the paying public necessitated careful selection of staff.

[12] Only insured persons, not their families as in the case of voluntary hospitals, are granted gratuitous care in the public hospitals on the basis of membership. This discrimination was devised purposely to promote use of the voluntary hospitals—to foster which the Hospital Savings Associations were started.

patients of the institutions. The council, however, has obligated itself in its adopted rules to the provision of no such limited service.

It is the opinion of more than one informed medical man familiar with current developments that both specialist services and institutional care administered by county councils inevitably will become insurance benefits. According to such opinion, the new benefits will be provided through public hospitals (and probably approved voluntary hospitals as well), through out-patient clinics, and through a domiciliary specialist service, the institution of which will become compulsory obligations of the county councils. The insured worker and his dependents will then receive these services as he now obtains general practitioner attendance as part of his insurance medical benefit, instead of being admitted as a Hospital Savings Association member without charge or paying according to the discretion of the almoner.

XI

SOME ALTERNATIVES TO THE HEALTH INSURANCE SERVICE SPONSORED BY MINORITY MEDICAL OPINION

OBVIOUSLY not all medical opinion endorses the exact health insurance views of the organized medical profession speaking through the British Medical Association. The most coherent minority medical group is embraced in the Medical Practitioners Union, an affiliate of the British Federation of Labor, i.e. the Trade Union Congress. Some of its membership is also found in the British Medical Association, but for the most part it comprises British Medical Association dissenters.

In its general approval of the health insurance system as a method of organizing the practice of medicine, it is at one with the larger organization. It differs with the latter only in desiring to develop from this organization of medical service a system for all persons of moderate incomes.

The Medical Practitioners Union advocates the immediate provision of general practitioner service, paid for out of the general tax receipts, for all families below a liberal income level. As in the health insurance scheme, all doctors could offer their services, free choice of practitioner by the patient would be allowed, and the government would pay the doctors on a capitation basis for the families on their medical lists. Specialist services similarly organized are advocated for addition after five years' operation of the general practitioner service.

The Union officials point out that this arrangement would immediately extend the advantages of the health insurance medical service to shopkeepers, farmers, and similar groups whose needs are just as pressing as those of the technical wage earners who are the only group that with administrative ease can be included in a contributory insurance system. "State" doctors, it is emphasized, would perform all the duties now assigned to the panel doctors. No substantial change, therefore, would be called for in the organization now operative for the supervision of medical benefit. Health insurance as such would continue to provide the wage earners during incapacity with cash benefits which would be certified as usual by the attending physician.

This proposal is violently opposed by certain other minority medical opinion, which is pressing for a complete state or socialized medical service along lines often discussed by advocates of "state medicine" in the United States. It would be inaccurate to speak of the medical men who are of this opinion as a group comparable to the Practitioners' Union, for example; they are not comparably organized and there are many individual differences on various points involved. The chief basis of their opposition to the Medical Union's plans and those of the British Medical Association lies in the very principles upon which the British Medical Association and the Medical Union agree, i.e. the right of all doctors to participate, completely free choice of doctors, and capitation. This "state medicine" group deems the only desirable medical service to be one with selected, salaried staffs serving all the public as does the free school system. A few of them think that the whole "choice of doctor" theory is much ado about nothing. Choice by the patient within the selected staffs is conceded by most of them.

This point of view, critical of open panel and free choice, is well expressed in the following excerpts from two papers written by Dr. Sommerville Hastings, one of its exponents, who is both an eminent surgeon and teacher and an important voice in the London County Council Medical Service:[1]

[1] *The Evolution of a State Medical Service*, by Sommerville Hastings, M.S., F.R.C.S., surgeon and lecturer, Ear and Throat Department, Middlesex Hospital; chairman,

"It [extension of insurance services] would result in any practitioner within certain wide limits, whatever his record or qualifications, who could curry favor with a section of the populace being guaranteed a living at the expense of the state. After all, it is very difficult for the public to estimate correctly the relative value of the services rendered to them and impossible in most cases for them to distinguish between the efficient and the inefficient doctor. Moreover, by maintaining the complete separation between the doctors who prevent and those who treat disease, it would accentuate one of the greatest evils of our present régime."

Again: "You will tell me that English people do not like being regimented, that they will not put up with the doctor provided for them by the State, but will want to choose their own. I quite agree. But there is no reason why a patient should not choose his own doctor in a State medical service, although obviously a wider selection will be possible in town than in the country, just as it is today. Unless a doctor allows himself to be overworked and therefore inefficient, he cannot deal with more than a certain number of patients, and this applies equally to the doctor in private practice and in the national service. By allocating the required number of doctors to a given area, it would be easy to permit any individual to select his own doctor, just as is now carried out under the panel system. Doubtless the very popular doctor would have his full quota of patients and even a waiting list, but, on the other hand, there would be many people who would not trouble to make a selection, and these could be allocated to the less popular or more recently appointed doctors."[2]

None of the proponents of salaried, selected general practitioner service have in mind, of course, any sudden inter-

Hospitals and Medical Services Committee, London County Council. Reprinted from *The Medical Officer*, March 16, 1935, London: Sir Joseph Causton & Sons, Ltd., pp. 5, 6.

[2] Reprinted from *St. Bartholomew's Hospital Journal*, June 1936, London: Allard & Son, pp. 5, 6.

ruption of health insurance service. A gradual evolution is proposed so designed that while protecting the doctors who already have interests involved in panel practice, it will shut off future entrance to the panels. Salaried medical men, then, at first will supplement, and in the course of time gradually will take over the general practitioner care of the sick, insured as well as the uninsured.

Meanwhile, they propose to develop specialist services, both institutional and non-institutional, in connection with public (County Council) hospitals which they believe should be made obligatory at once. The likelihood that this last proposal may become a fact is agreed to by many who are not advocates of state medicine. In so concluding, they point to the long failure of the government to add specialists' services to medical benefit; to the great advantage, if not the necessity, of the laboratory and other facilities which connection with the hospitals gives a consultant service; and to the effect of the current London County Council developments.

Nothing which could be termed in any sense "medical opinion" in Britain today contemplates as either possible or desirable for any but the well-to-do an unorganized medical service arranged for by individual contract between patient and doctor.

PART II: DENMARK

XII

SUMMARY OF THE DANISH HEALTH INSURANCE SCHEME

THE Danish health insurance scheme dates back forty-six years to an act passed in 1892. This measure brought into existence a voluntary subsidized insurance scheme based upon the principle of *inducing* rather than *compelling* self-help. Mutual aid societies[1] were offered several types of government subsidy toward the cost of their health insurance benefits, on condition that they registered for government supervision, met certain requirements in their membership policy, and offered benefits of prescribed *minimum standards*.

These subsidies were to be available on behalf of only the "unpropertied" members, defined as either industrial wage earners or other persons, such as clerks, farmers, etc., with incomes and accumulations below stipulated amounts.

The government supervision was directed at (1) compliance with the benefit requirements; (2) ensuring financial soundness; and (3) guaranteeing membership to anyone in passable health who desired it. In other respects the societies were allowed to manage their own affairs as they chose without interference from the government. Persons suffering from incurable or chronic diseases were to be admitted only if the rules of a society permitted it, and many did. Such members were not to be entitled to benefit during periods of the special handicapping disease but were to be insured against all other types of illness.

[1] i.e. fraternal organizations like the English Friendly Societies of which there had been a considerable development following the dissolution of the compulsory gilds in 1862.

In promoting health insurance the Danish act, like the scheme in England, was directed at the protection of persons who were dependent upon the income they could earn by their personal labor. However, unlike the compulsory British system, which for administrative reasons was restricted to wage earners, the Danish act provided that small farmers, shopkeepers and other self-employed workers were equally eligible with the white-collar wage earner. With the voluntary subsidized principle, there was, of course, no reason why they should not be.[2]

The necessary minimum benefits were to be granted for thirteen weeks in the year and included medical care and hospital treatment for the member and his children under fifteen years of age and a cash benefit.

Other benefits, such as free medicine, massage, medicinal baths, home nursing, etc., were permitted but not required. Societies were permitted to deprive a member of cash benefit in case of illness resulting from intoxication or a brawl "originating with the member." Members paid varying premiums or dues according to the amount of cash benefit desired. The wife of an insured person was expected to be insured in her own right and did not derive any direct benefit from her husband's insurance.

Medical, hospital, and cash benefits to needy society members for a period equal to the insurance period, i.e. another thirteen weeks, as well as payment of dues during unemployment periods, were guaranteed at the expense of the commune without the usual penalizing effect of loss of civil rights which was imposed upon "poor relief" recipients.[3]

In contrast to the British and other compulsory health insurance schemes, the employer as such paid no part of the cost in this voluntary subsidized set-up. It was not uncom-

[2] See pp. 3 and 4 *supra* for explanation of coverage restrictions of compulsory schemes.

[3] Receipt of poor relief ordinarily meant loss of certain civil rights, e.g. loss of the right to vote and hold office, loss of old-age pension rights, loss of right to marry without communal consent, etc.

mon, however, for employers and other persons of means as become honorary members of a Registered Society, and to such to make an annual contribution to the society's treasury.

It is significant that medical and hospital care for the member's children, as well as for the member, were required. Since no regular share of the premiums was met by the employers, this would have meant prohibitively high premiums even with the government subsidies, had it not been for the nature of Denmark's hospital system. The hospitals of Denmark were largely governmental and derived their main support from taxes rather than from patients. In other words, to quote from an official description of Denmark's hospital situation, the public hospital system was, even aside from the insurance scheme, "a kind of insurance for the whole population."[4] It is as traditional in Denmark that hospitalization should be financed mainly by public funds as it is in the United States that public schools should be entirely supported by public taxes and available to rich and poor alike. Hospital care, moreover, as in Britain, includes attention from the medical and surgical staff as well as maintenance and nursing.

In twenty years' time, i.e. by 1913, when the British system was just getting into full swing, almost two-fifths of Denmark's population were receiving both medical and hospital care through her health insurance scheme.

By this date also, the Copenhagen societies were providing specialists' services as well as family doctor attendance. Most of the urban organizations and some of the rural ones were paying all or part of the cost of drugs and surgical appliances. Important changes occurred in 1921. One amendment required societies to admit physically handicapped persons to full membership and unconditional benefit if such persons were able to make any contribution toward self-support. Special subsidies reimbursed the societies for three-

[4] See *The Danish National Insurance Act*, by Sigurd Wechselmann, published under the auspices of the Ministry for Social Affairs, Copenhagen, 1936, p. 27.

fourths of the additional expenses entailed by the illness of such members. (The commune was obligated to reimburse the national exchequer in the amount of one-half of these special subsidies.) Another amendment required maternity benefits, which included both cash allowances for ten days after childbirth and medical care, when necessary, during confinement.

Compulsory invalidity insurance for all "unpropertied" members insured for health insurance was legislated in this same year, i.e. 1921. Employers were required to contribute to the cost of this insurance by payments which were collected with their workmen's compensation premiums for industrial accident insurance.

To date, however, Denmark has remained unwilling to abandon completely her "inducing self-help" principle in the social security field, as was demonstrated in the two omnibus statutes passed in 1933, the Social Security Measure and the Public Assistance Law. These enactments worked a revision of Denmark's social welfare set-up. Compulsory invalidity insurance was made universal.[5] Health insurance became a complicated hybrid of the compulsory and voluntary principles of social insurance with an admixture of a "preferred" type of relief.

The old-age pension system is still non-contributory. No old-age pension, however, can ever be obtained by a person who evades his legal obligations with respect to health insurance.

Voluntary health insurance societies are still to be Registered Societies and still to receive subsidies on account of their "unpropertied" active members. In addition, non-subsidized Health Insurance Associations "Continuation Funds" for the better-to-do, although not subsidized, are to be "state inspected," i.e. subject to state supervision and control.

[5] The government pays about one-fifth of the costs of this insurance, and no pension is given to the "undeserving" person who persistently leads a flagrantly immoral life.

It is now, however, incumbent upon every person between the ages of twenty-one and sixty, unless already insured against sickness, to seek either active ("participating") or passive ("contributing") membership in a Registered Society or in a Health Insurance Association. If able to make any contribution toward self-support and not actually ill when application is made, membership cannot be refused. "Propertied" persons who desire active "participating" membership must join either a state inspected Health Insurance Association or a special section of a Registered Society.[6]

Passive "contributing" membership may be taken with either type of organization by the well-to-do person, and with a Registered Society by the man or woman without means (i.e. "unpropertied").

A passive "contributing" member pays a small annual[7] sum as a premium and *no benefits* are payable to him. He has, however, two valuable rights: (1) at any time in his life up to the age of sixty years, should he become an "unpropertied" person, his status may be changed to that of active participating membership with full benefit rights; and (2) at any time before the age of forty years, even though he remains "propertied" and so not eligible to participating membership in a Registered Society, he may demand active membership in a Health Insurance Association.

There are eighteen Health Insurance Associations covering all parts of Denmark. Persons who, at any time up to the age of sixty years, pass out of participating membership in the Registered Society because of improved income may become participating members of a Health Insurance Associa-

[6] A Registered Health Insurance Society already having a separate section for "propertied" members may, if it has not less than 15 members, continue this section on the same basis as the Health Insurance Associations, i.e. without public subsidy.

[7] Two kroner (54 cents) until twenty-five years of age; 2.50 kroner (67½ cents) thereafter. This must be collected from what Danish official reports term the "reluctant" contributor either with the collection of local taxes, which almost every Dane is required to pay, or by separate process in the case of the non-taxpayers. Krone figured at 27 cents.

tion.[8] These associations do not directly provide medical care, but partially reimburse their members for their medical expenses.

In brief, the plan set up by the 1933 legislation is that every Danish citizen of appropriate years shall become at least *potentially* insured against sickness and shall pay a specific annual sum for this privilege.

The government fixes each year, by reference to statistical data on wages and the cost of living, the income limit and the property limit set up for determination of "unpropertied," as distinguished from "propertied" persons. The objective is that the limits of income fixed for the "unpropertied" group shall correspond with the annual earnings of the skilled worker working full time and that accumulation of a reasonable amount of property shall be encouraged. The general rules for such determination may be revised every third year. In 1936 the *income* limit was set at 4,200 kroner ($1,134) in Copenhagen, 3,600 kroner ($972) in the towns, and 2,800 kroner ($756) in the strictly rural areas. A maximum figure for *property* owned was set for the entire kingdom of 14,000 kroner ($3,780) in the case of breadwinners and 9,500 kroner $2,565)[9] for other persons.

The main outlines of the present health insurance provisions for "*unpropertied*" persons are as follows: The qualifying period for benefits is the very short one of six weeks, and even that period is dispensed with in event of accidental injury and also in the case of persons under fifteen years of age who are insured in their own right. Maternity benefit is due only after ten months' insurance.

Medical benefit[10] remains practically as set up in the original act.

[8] This privilege of transfer, which is permitted those who have been actively insured, is justifiably more extensive than that allowed the merely passive or contributing members, who, as has been stated, must transfer to the Health Insurance Association before the age of forty.

[9] Krone at 27 cents as stated in n. 7 *supra*.

[10] The Workmen's Compensation Act entitles the worker who suffers an industrial

Cash benefit must not be paid for the first three days of illness, and societies are permitted to set a *seven*-day waiting period. No cash benefit is payable in case of a compensated industrial injury, although regular medical benefit is granted. Further, no cash benefit is ever payable unless income during illness is either cut off or reduced. Lastly, cash benefit, when added to such income (except when an invalidity or old-age pension is the sole source of support), must never be sufficient to total more than the amount of customary income.

Daily cash benefits ordinarily must not be less than forty öre (the price of two quarts of milk in Copenhagen), nor more than six kroner (30 per cent more than the farm laborer's wage). In no event must they exceed 80 per cent of the insured person's customary earnings. Women and minors under eighteen may be permitted to insure for a lower amount or to dispense with cash benefit altogether. The society's rules may provide that cash benefit be reduced or suspended during hospitalization, and benefit still may be denied altogether in case of disability resulting from intoxication or a "brawl" precipitated by the claimant.

Cash benefits[11] must be provided for twenty-six weeks in the year up to a maximum of sixty weeks in a three-year period. The rules permit two thirteen-week benefit periods separated by an interval of thirteen weeks, and this is the usual arrangement. During the thirteen-week intervals, up to a similar maximum of sixty weeks of benefit in three years, the commune is obligated to provide identical "continuation" benefits to all needy insured persons. This is not deemed ordinary poor relief, but a quasi-insurance right. The member who draws sixty weeks' benefit (which is his maximum)

injury to additional special medical services which are not furnished by his Health Insurance Society, and to artificial limbs, glasses, orthopedic appliances, etc., as well as to cash compensation.

[11] The time limits for medical benefit are the same as for cash benefit when the doctors are paid on a fee basis. See pp. 140*ff*.

in three successive years, has his active status suspended for at least a year.

Reinstatement as an active or participating member after suspension rests upon the same physical qualifications as original admission, i.e. being able to make *some* contribution toward self-support. If unable to make such a contribution and therefore inadmissible to sickness insurance, the applicant is undoubtedly eligible for an invalidity pension which is payable on loss of two-thirds of his earning capacity. This carries with it the right to necessary medical and hospital attention at the expense of the commune,[12] as is equally the case with an old-age pensioner who loses his sickness insurance membership.

The maternity benefits provide a cash allowance equal to the regular sickness benefit allowance and payable for a fourteen-day period after confinement. (Free midwife assistance and medical attendance are also guaranteed.) Women who are required by the Factory Act to abstain from work for four weeks after childbirth receive a further cash allowance equal to about half the minimum factory wage. The allowance is continued for another two weeks, moreover, if the mother remains away from her work for this additional period in order to nurse her child.

The same cash allowance must be paid for eight weeks before confinement to women in those listed industries in which work is deemed especially detrimental at such a stage of pregnancy.

Insured persons who move from one area to another are transferred to the society in the new area with all accrued membership rights. This arrangement has been effective for some years and was worked out by voluntary cooperation of the central council of Registered Societies and the Directorate of Health Insurance Societies, the government's supervisory office. Societies are grouped into District Federations for the purpose of making arrangements for medical

[12] Whether a "means test" may be exacted is in dispute, see Chap. XVI, n. 2.

care and their arrangements must be approved by the Minister of Social Affairs. District federations are in turn members of a Central Union covering the entire country. This Central Union has cooperative agreements with corresponding Unions in Norway and Sweden which provide *inter alia* for transfer of insured persons from societies of one of the Scandinavian countries to societies functioning in one of the two others.

The direct governmental subsidies amount to more than 40 per cent of the members' dues. No subsidy is granted in respect of administrative costs or of expenditures on drugs. The hospitalization costs met by the commune over and above what is paid by the societies for treatment of their members more than double the government's actual contribution. The communes may not charge societies for hospitalization of their members more than 50 per cent of their usual rates. Since the latter constitute at most only one-fifth, and often a smaller fraction, of the actual hospital costs, at least 90 per cent of the institutionalized insurance cases are financed directly by public taxes.

In 1936, there were 1,609 societies registered for health insurance; 650 of these had a membership of less than 500. The largest organization (in Copenhagen) had about 240,000 members. The average active membership was 1,309; including "contributing" membership it was 1,433.

Altogether, 2,100,000 "unpropertied" persons over fifteen years of age were insured. This constituted 77 per cent of that age group. With their 800,000 children, who also were entitled to medical and hospital care, they constituted approximately 88 per cent of the Danish population, in contrast to the 40 per cent of the population receiving insurance medical benefit in England. The average weekly contribution paid by the insured member was 0.41 of a krone, about 11 cents.

The Health Insurance Associations for the better-to-do had 112,000 members and the State Railway Health Insurance Association some 38,000 more.

While the average cash benefit is considerably higher in the Danish metropolitan area than the English benefit,[13] this benefit is, like the English standard benefit, often far too low for family maintenance. The Directorate of Health Insurance Societies believes adequate cash allowances to be of great importance from a health as well as an economic standpoint and constantly urges the societies to more intensive campaigns for larger money benefits, especially in the rural areas. This, of course, means inducing the insured to pay larger contributions, as dues vary with the amount of cash benefit desired.

The most significant feature of the Danish system has always been its emphasis upon medical and hospital care. A continually increasing percentage of total expenditures has been devoted to such benefits in comparison with the cash benefits paid.

Arrears because of unemployment or other calamity need not interrupt the insured person's protection, as the commune is obligated to keep up dues when the insured, because of such circumstances, cannot afford to pay them.[14] This situation, like the similar one in Britain,[15] not only protects the working family but also brings to the Danish physician the security of continued income in bad times as in good.

[13] i.e. for adult males. Eighteen kroner ($4.86) per week in comparison with the English standard of fifteen shillings ($3.75).

[14] Dues are paid to insure medical and other service benefits and cash benefit at a rate of not more than two kroner per day.

[15] See pp. 8 and 9 *supra*.

XIII

HEALTH INSURANCE MEDICAL BENEFIT

THE medical benefit required by Denmark's health insurance system is available immediately on taking out insurance and provides a "family doctor" or general practitioner service as does the British scheme. In contrast to British medical benefit, which is available only to the wage earner, both the insured person and his (or her) children under fifteen years of age are entitled to such service in Denmark. "Children" includes stepchildren, adopted children, and foster children who are placed "in the home to be reared." In addition, and again in contrast to the British scheme, medical benefit provides hospital care (including treatment in institutions for mental cases) as well as sanatorium treatment in case of tuberculosis. Instead, however, of the generous provision of all "necessary drugs and medicaments" which is provided in England, payment of three-fourths of the cost of insulin for treatment of diabetes, liver and similar compounds for pernicious anemia, and eucortone (a hormone for Addison's Disease), is all that *must* be done for the insured person's drug needs. Except in the metropolitan area (i.e. Copenhagen and its environs) free transportation to and from the office of the physician and the hospital must also be furnished as an insurance right, but by the *commune* not by the sick fund.

Although it is unlawful, as in England, for the doctor to require payment from the insured patient, many societies through the "control ticket" system, collect small amounts from members who seek medical attention. The organizations that pay their doctor chiefly on a capitation basis—organizations which include in their membership about three-fifths of the total number insured—quite commonly require the insured person to use a "control ticket" or to get the doctor's endorsement of a control card when obtaining specially paid-for medical service. Such service usually includes calls made by the doctor during the night or on Sunday or a legal holiday. The usual fee collected by the town and rural society for each control ticket or endorsement of the control card is fifty öre (13½ cents) or one krone (27 cents), while two kroner (57 cents) to four kroner ($1.08) is collected in the metropolitan area, i.e. Copenhagen.

Many of the rural organizations which pay their doctors on a fee basis require a control ticket (or an endorsement of a control card) before consulting the doctor *at any time*. As a rule a smaller sum is charged for this control service when the patient goes to the doctor's office instead of asking the doctor to make a call at the patient's home. Frequently the control fee for a follow-up call made at the doctor's initiative is less than that required when the patient sends for the doctor. The charge is generally doubled when medical service is demanded at night or on Sundays and holidays. About one-tenth (120) of the rural societies which use the control system place a maximum on the number of control tickets or endorsements which may be demanded in any one illness.

In 1935 the total amount collected in "control" fees by all the societies in Denmark was 776,000 kroner ($222,800), which amounted to *between 3 and 4 per cent* of the amount paid for remuneration of the health insurance doctors.[1]

As in England's health insurance scheme, childbirth is not regarded as illness, and the general medical service provided does not of itself include obstetrical care. Necessary medical attendance at confinement and delivery is included in the insurance, however, as a separate maternity benefit. Such attendance is deemed "necessary" if in the opinion of the

[1] Information furnished by Direktor Borberg of the Danish Health Insurance Inspectorate.

insured woman's midwife or doctor such medical aid is required. Midwife and nursing assistance for the mother and child and the necessary dressing, supplies, etc.,[2] must also be provided by the health insurance society.

In 1934 maternity benefits were given by the health insurance societies in connection with almost three-fourths of the births in Denmark that year.[3] Medical assistance was furnished in a little under 30 per cent of the insurance confinement cases.

The optional part of medical benefit plays a most important rôle in the Danish scheme. All the societies in Copenhagen and almost two-thirds of those outside the metropolitan area, pay *part* of the cost of all medicines prescribed by the attending physician. As the law requires the insured to pay at least one-fourth of any drugs provided under the insurance scheme, there is no completely free drug service. The Copenhagen societies, those in the smaller cities, and a few of the rural funds pay the maximum permitted amount, i.e. three-fourths of the cost. Most other organizations meet one-half of the drug costs of the insured and a very small group meets less than half.[4]

All urban societies and most of the rural ones provide a dental benefit, consisting of a contribution toward the cost of specified dental care. In the Copenhagen district the specified care is defined as "all necessary treatment of the teeth and oral cavity," except gold fillings and artificial dentures. Treatment of diseased gums, as well as all work necessary to the preservation of the teeth, is included.

A third of the organizations outside of the Copenhagen area offer similarly broad dental benefit and twice as many (more than a thousand) other funds, pay for extraction of teeth. The latter service, however, is frequently performed

[2] The Public Health Service stipulates the standard amounts that must be furnished.

[3] 73 per cent of all births. Wechselmann, *op. cit.*, p. 47.

[4] The Copenhagen societies (43); the other urban societies (75); and eight rural organizations pay three-fourths; 800 organizations pay one-half, and 59 pay less than one-half of the cost of the medicines.

by the attending doctor, and if special dental benefit is not provided, extraction of teeth and treatment of gums is a medical obligation.

Except when extraction only is provided, the insured person in all cases is required to pay a charge directly to the dentist. This ranges from 20 per cent to 50 per cent of the total cost of the dental care. The charges for insurance dental care are regulated by contracts between the individual dentists and the societies, which embody the terms of an agreement drawn up by the Dental Association and the Federated Societies.[5]

All societies in the metropolitan area of Copenhagen provide specialists' services covering all the important broad fields of specialization. These fields include:

1. Ear, nose, and throat.
2. Eye.
3. Lungs, kidney, and heart.
4. Pediatrics.
5. Dermatology and syphilology.
6. Orthopedics.
7. Neurology and psychiatry.
8. Gynecology.
9. Obstetrics.
10. Roentgenology (X-Ray).
11. Gastro-enterology.
12. Specialized surgery (brain, abdomen, rectal, etc.).
13. "Massage,"[6] including corrective gymnastics, diathermy, hydrotherapy, violet ray, etc.

Treatment by orthopedists, dermatologists, and also by "massage" specialists is granted by the organizations in some of the towns and rural areas. But, except as given through the out-patient clinics of the hospitals, regular specialist

[5] This contract follows the general form of the Medical Contracts.

[6] For explanation of this term, see p. 121 *infra*.

services outside of the Copenhagen area, are confined usually to eye, ear, nose, and throat work. Eight hundred societies have agreements covering the latter services. No official record, however, is kept of the extent to which these services are made actually available to the insured. It is believed by the government supervisory officials[7] that at least the seventy-five societies in the towns, as contrasted with the funds in the purely rural areas, give their insured members all needed care in these special fields.

Denmark is divided into twenty-four medical districts, in more than half of which there is now at least one "central" hospital which has modern technical equipment and a fairly standardized specialist staff in both surgery and internal medicine. The rural and town health insurance societies arrange in the more serious cases both hospitalization and free transportation to and from these "central" institutions (instead of the nearest communal hospitals) and in many cases also pay for treatment in their out-patient departments.[8]

Practically all of the societies in districts that are still without a central hospital pay part of the specialist fees in certain urgent cases.

Two-fifths of the organizations (650) add home nursing to the medical benefit which their members are entitled to receive, and in Copenhagen and a few of the towns the Federated Societies have established convalescent homes.

As in Britain, the health insurance doctor is primarily responsible for deciding what drugs to prescribe for his insured patient and is similarly obliged to observe rules of economy when several preparations are equally efficacious. As has been remarked, the patient must pay at least one-fourth of the cost of the drugs, a rule that is itself directed against extravagant use of medicaments.

[7] Information furnished author by Direktor Borberg of the Danish Health Insurance Inspectorate, August 1936.

[8] The cost of transportation to the central hospital, when provided, is divided among the society, the commune, and the county. See Leg. Ser. 1933, Den. 4, No. 17 (4).

If the insured person persistently refuses to follow the instructions of his physician or refuses to accept hospitalization when his doctor deems it necessary, he forfeits all right to medical benefit in Copenhagen and the right to anything but treatment at the doctor's office in most of the non-metropolitan area.[9] The doctor must, of course, notify the society when such a situation develops.

As is the case in England, more is expected of the family doctor practising in the country than of the urban general practitioner.[10] This principle is recognized in the district collective contracts by which the terms of service of the Danish health insurance doctor are determined.

Disputes about the limits of the doctors' obligations are, as in Britain, settled by a procedure in which the organized medical profession plays an important part. Cases are referred first to the District Medical Association and the District Federation of Societies. If these two organizations cannot negotiate an agreed decision, the question then goes to an arbitration council or board[11] composed of an equal number of representatives from each body. Their decision is final.

Again, as in England, the doctors' obligations include the giving of certificates of capacity and incapacity to guide the health insurance society in administering the cash benefit to which the insured worker is entitled during incapacitating illness.

Medical benefit may be suspended normally when a member is in arrears for a period stipulated in the society's by-laws (as a rule two or three months). Extensions are usually granted on request. In addition, since the commune is ob-

[9] cf. the British rule and see p. 12 *supra*.

[10] Obviously, when the society gives a special dental benefit as one of its optional benefits, a dentist rather than the doctor extracts teeth.

[11] In Copenhagen this is a special local board. Outside Copenhagen, a central Arbitration Council serves for the whole kingdom. The Council is composed of a chairman and six members, three elected by the Danish Medical Association and three by the Federation of Health Insurance Societies; the chairman elected by these if four agree, otherwise appointed by the Minister of Social Affairs. See pp. 150*ff*.

ligated, as has been stated,[12] to pay the dues of any insured person whose financial difficulties are caused by unemployment, illness, or any other circumstances over which he has no control, penalized arrears are inflicted only upon the "undeserving." The latter periodically pass out of the insured group to become the problem of the poor-law authorities.

In Denmark in the year 1935, the following expenditures[13] were made on health insurance benefits for insured persons.

HEALTH INSURANCE BENEFITS, 1935

	kroner	
Cash benefits	10,611,331	($2,865,053)
Remuneration of general practitioners	17,762,368	($4,795,839)
Specialists	2,658,928	($ 717,910)
*Institutional and clinic treatment	12,662,775	($3,418,949)
Drug benefit (voluntary)	6,736,699	($1,818,908)
(required)	399,315	($ 107,815)
Surgical appliances and glasses	874,748	($ 236,181)
Maternity benefit	2,652,515	($ 716,179)
Home nursing	1,040,988	($ 281,066)
†Transportation of patients	36,252	($ 9,788)
Physiotherapy massage (by a physiotherapist not physician)	374,198	($ 101,033)
Medicinal baths (other than hospital cases)	160,554	($ 43,349)
Dental benefit	2,867,309	($ 774,173)
TOTAL	58,837,980	($15,886,243)

*Only a fraction of the cost of such care; see p. 109 *supra*.

†In connection with transportation to central hospitals and not to be confused with transportation expenses met by the communes; see p. 115 *supra*.

The major part of the communal hospital service is, of course, rendered on behalf of insured patients. Therefore, as has been stated previously, most of the outlay of the communes on its hospitals over and above that which is paid by

[12] See p. 110 *supra*.

[13] *Beretning fra Direktoren for Sygekassevaesenet*, I Aaret 1935 (Report of the Directorate of the Sickness Insurance Funds for the year 1935), pp. 22, 23. Krone figured at 27 cents.

the societies for hospitalization of its members is really a public expenditure for the institutional treatment of insured persons. This adds approximately 25,500,000[14] kroner ($6,-885,000) more to the expenditures made for health care of insured persons.

[14] See Wechselmann, *op. cit.*, p. 46.

XIV

THE HEALTH INSURANCE DOCTOR IN COPENHAGEN

TO be a health insurance doctor in Denmark, the practitioner must be a member of the Danish Medical Association as well as willing to work on the terms offered by health insurance practice. Theoretically, a non-organization physician, by securing permission of the Association, may enter health insurance practice. Practically, this does not happen, as such permission is not given.

The details of the terms of insurance practice are settled in collective agreements entered into between either the county or district (i.e. a group of counties) Medical Associations and the corresponding county or district Federations of Health Insurance Societies. To be effective, agreements must be approved by the Central Executive Council of the Danish Medical Association and by the Minister of Social Affairs. The latter, before giving his approval, consults both the Central Council of the Federated Health Insurance Societies and also the Director of the National Health Service, which is the Department of Public Health in Denmark. In case no agreement is arrived at, provision is made for submission of points of difference to a Central Arbitration Council[1] if both parties so desire. In the event of a deadlock which one side is unwilling to arbitrate, the Minister of Social Affairs attempts to bring the parties to a settlement. Finally, should there be hopeless disagreement, the societies must arrange that their members pay their own medical bills with the right to partial reimbursement (not more than three-fourths of the cost) from their organizations. It is over

[1] See p. 116, n. 11, for details of this Council.

fifteen years since such a deadlock occurred in any district in Denmark, and it is believed in medical circles that there never will be another one.

There are seven district agreements effective at present in the following broad areas: (1) Copenhagen and environs; (2) Zealand (outside the Copenhagen area); (3) Jutland; (4) Bornholm; (5) Laaland; (6) Fünen; (7) The Faroe Islands.

In the Copenhagen area a limited list, which includes 71 per cent of the doctors in general practice, provides the required family doctor service for the insured population. In the rest of Denmark, it is the general rule that any physician who chooses to do so may be a health insurance general practitioner and practically all the doctors have some health insurance practice. The provision of a limited panel in Copenhagen has full support of the Medical Association, which believes it a wholesome deterrent to overcrowding the profession in the metropolitan area.

As the conditions in Copenhagen are different from those prevailing elsewhere, Copenhagen will be discussed separately. This will be followed by a similar survey of the situation in the rest of Denmark with special reference to the Jutland collective contract as illustrative of the prevailing rules in the non-metropolitan area.

The Copenhagen Area

As has been stated previously, all Health Insurance Societies in Copenhagen provide complete specialist service as well as the required family doctor treatment. The Federated Health Insurance Societies of Copenhagen enter into two agreements with the local Medical Association to regulate the position of the health insurance doctors. The present agreements were signed in 1936 and are binding for ten years (Collective Contract I) and five years (Collective Contract II) respectively.

Collective Contract I covers the doctors who are "regularly employed." It includes both general practitioners and also three types of specialists (1) eye, (2) ear, nose, and throat, and (3) "massage." This last is a specialization more elaborate than the physiotherapy of American medical practice. It combines corrective manipulation, diathermy, hydrotherapy, and corrective gymnastics. Most of the massage specialists are connected with "massage clinics" owned by the Federated Societies. These doctors handle four-fifths of the insured patients who need this special service. Pending the availability of adequate Federation-owned clinics, additional specialists in the corrective massage field, who have their own private clinics, are employed.

Collective Contract II covers *occasional* service, and includes physicians in the following special fields: (1) specialized surgery, i.e. brain, abdominal, rectal, (2) gynecology, (3) obstetrics, (4) dermatology and syphilology, (5) gastroenterology, (6) orthopedics, (7) pediatrics, (8) heart, kidney, and lungs, (9) neurology and psychiatry, (10) roentgenology.

The doctors practising under the first agreement, i.e. the "regularly employed" doctors, included in 1935, 252 general practitioners, 21 eye specialists, 27 ear, nose, and throat men, and 10 "massage" specialists.

When appointments are to be made to this medical panel, the Copenhagen Medical Association must be notified of prospective vacancies at least three weeks in advance of the time set for receiving applications. The Medical Association publishes notices of such vacancies in the *Medical Journal.* A Joint Committee of Six,[2] comprising three doctors appointed by the Danish Medical Association (one of whom must be a specialist when the vacancy is in the specialist group), and three representatives selected by the Federation of Societies, then considers applications and makes recommendations to the latter Federation's business committee. The Federation appoints the doctors. This method of appointing the medical panel has not always been in operation in Copenhagen. Until 1921 *all* of the societies, and until 1926 *some* of the societies

[2] Committee on Vacancies and Appointments.

directly employed their own physicians. Since 1926 all further appointments have been made through the Federation. Of course certain physicians are still practising under appointments dating back to the period when the individual society made its own arrangements for medical care. Only 113 of the 252 health insurance doctors in Copenhagen today are employed by the Federation to serve *all* societies.

Physicians performing the "occasional service" specialist services regulated by Collective Contract II must belong to the official Specialists' Organization, recognized by the Medical Association of Copenhagen. Specialists who wish to do this "occasional service" health insurance work must, before the end of September, send the Medical Association details of their office hours and consultation rooms. The Medical Association then sends the Federation of Societies by October 15 a list of available specialists. Before January 1 the two organizations send out a joint notice to each society and to each general practitioner listing the specialists to whom the insured patients can be referred during the ensuing year.[3]

The area is divided for "family doctor" service into fourteen general districts (with subdistricts) and the general practitioners are assigned to the various districts. The fourteen districts average eighteen doctors to the district, roughly one doctor to each 1,500 insured. The largest number of physicians in any one district is twenty-eight, the smallest is nine. The agreement requires that at least one woman shall be selected for each district when possible, and there is one woman on the panel of health insurance doctors in each of the districts.

The doctor has "preference" in the subdistrict of his residence. "Preference" means that insured persons are automatically assigned to the resident doctor of their sub-

[3] See pp. 130*ff.* for discussion of payment of these physicians as contrasted with the regularly employed eye, ear, nose, and throat, and "massage" specialists.

district unless they expressly register a desire to have another of the district doctors.

Choice among the doctors assigned to the district must be made by the insured persons for a year commencing January 1. Notification of desire to change doctors must be made during the month of December. The doctor's privilege of relieving himself of a patient, in the absence of agreement with his patient, must be similarly exercised, i.e. notification in December to be effective in January. If a patient is rejected by the doctor of his choice and asks to be assigned to another physician, ordinarily he is allocated to the district physician who was last added to the medical list. This practitioner must accept him.

With the permission of the physician in question, the insured person may select one of the practitioners in an adjacent district. A married couple who are living together are required to choose the same family doctor, who serves their children as well.

Societies with not more than 5,000 members are allowed to restrict the choice of physician of their members as follows:

In societies with a membership of
- not more than 2,000 members—choice of 2 physicians
- from 2,000 to 3,500 members—choice of 3 physicians
- from 3,500 to 5,000 members—choice of 5 physicians

As in Britain, the Danish doctor must arrange for a deputy to handle his health insurance practice when he cannot attend to it personally. If he does not customarily (i.e. even in private practice) make night calls, and most of these doctors do not, he does not have to do so for his insured patients. In such a case, the society furnishes at its expense, not at the expense of the regular doctors, "night watch" physicians whose function is to make night calls and report to the regular attending practitioners.

The doctors, whether general practitioners or specialists, are allowed to have deputies do their visiting on Sundays and holidays. The same rule applies to Christmas Eve, which is not an official holiday. All such visits are specially paid for by the societies, whether made by a regular doctor or by a substitute. A specialist may have as a deputy only a man in his own field. The doctor must notify the Federation and post at his consulting rooms full details of his deputy arrangements. If he plans to be away on vacation for more than two months or wishes, for any other purpose, to be absent from his practice for longer than three months, approval of the Federation is required. A deputy arrangement necessitated by illness is referred after eighteen months to the Joint Committee which passes upon vacancies in the medical list.

There is no rigid rule about limitation of medical lists, i.e. lists of potential patients. Instead, it has been provided for some time that the committee on vacancies at the end of any calendar year may in its discretion reduce (1) the list of a district physician that exceeds 2,000, (2) the list of an eye specialist that exceeds 25,000, (3) the list of an ear, nose and throat specialist which exceeds 20,000, and (4) the list of a "massage" specialist that exceeds 30,000. Not more than 500 names may be removed at any one time from the list of a district physician nor more than 5,000 from the list of a specialist. Moreover, insured persons who object to removal from their doctor's list through a "reduction" may retain their doctor by filing a request with their society.

Apparently, the popular doctor is given considerable leeway. In 1935, 90 of the 252 district physicians, i.e. general practitioners, had lists in excess of 2,000. Thirty of these doctors had more than 3,000 apiece. These lists mean, since the wife is listed separately from her husband, about half as many households. Hospitalization, moreover, is an insurance benefit and one which as a matter of public[4] policy, is generously employed by the practitioner. This removes to the medical staff of the hospital many cases which in England

[4] The Danish housing situation is a recognized problem. Because of congestion, it is deemed advisable to institutionalize many patients who with more adequate homes, could be handled well with home nursing.

remain the responsibility of the insurance doctor. One district doctor in Copenhagen had more than 4,000 on his list.

Similarly, five of the twenty-one oculists and six of the twenty-seven ear, nose, and throat men had lists substantially in excess of the possible limits.[5] As was emphasized in discussing the British health insurance doctor, these medical lists are, of course, *not lists of actual patients* but of persons who look to the doctors in question for medical care if and when they need it.

It should be remarked also that insurance doctors do not attend invalidity pensioners as do the British insurance practitioners.

The district physician is required to have on every week day at least one regular consultation hour, which may not be varied[6] for the different days without special permission of the committee on vacancies. For the convenience of the insured whose employment makes the usual office hour unsuitable, the doctor must be available for consultation by such persons at least one hour per week after 6:00 p.m. Rules specifically require the waiting room to be accessible a quarter of an hour in advance of appointments.

In seeking medical care the patient is expected, except in emergencies caused by accidents or sudden illness, to identify himself as an insured member in good standing by presenting his membership book or a medical card from his society. The doctor is entitled to charge for his services if such identification is carelessly omitted.

Again, except in case of emergency, the patient is supposed to request medical calls between 8:00 and 9:00 in the morning. He must, of course, consult the doctor at his office whenever it is physically possible, and he is specially instructed not to call the doctor during the night or on Sundays and holidays, except in cases of "absolute necessity."

[5] One ear, nose, and throat man had practically 50,000 "potential" patients and one of the eye men had a list of over 43,000.

[6] Except Christmas Eve from 9:00 a.m. to 4:00 p.m.

A person entitled to insured medical service who falls ill while away from home must be given care through the local society in that region. This society is entitled to reimbursement from the patient's own health insurance organization.

It is specifically stipulated in the Copenhagen Collective Contract that the general practitioner's obligations do not include making laboratory analyses, giving inoculations, extracting teeth, or performing operations other than those which he can properly conduct at his office, without the assistance of a second physician. The latter proviso automatically excludes operations which require a general anesthetic. The insured may obtain the excluded services as part of medical benefit through a specialist on the health insurance panel, in the hospitals, in the outpatient departments of the hospital, or through the dental service which all Copenhagen societies provide.

The patient, on pain of possible loss of medical benefit, is obligated to follow the instructions of his doctor and to enter the hospital for surgical or other treatment when the doctor so advises. He is not required, however, to undergo treatment which may hazard his life or his health. On disobedience of his instructions, the doctor reports the matter to the insured person's society. If, after discussion with the society committee, the patient persists in his refusal to obey medical orders (including the direction that the patient undergo an operation), the physician is released from obligation to treat him further.

The society then either transfers the patient to another physician with the permission of the two doctors involved, or makes special medical arrangements or, if the patient's action is not legally justified, suspends medical benefit. In the last case the insured, until willing to follow directions, is left to make private arrangements for medical care at his own expense.

This is somewhat in contrast to the British rule which, it will be remembered, similarly requires the patient to obey medical instructions, but specifically states that, except in case of a minor operation unreasonably refused, refusal to submit to a surgical operation or *to vaccination* or *inoculation of any kind* shall not subject the insured person to penalty or suspension of medical benefit.

Specialist service is available as part of insurance medical benefit when and to the extent that the *family doctor* deems it necessary.

The general practitioner is regarded as the key person in the medical set-up. He has the ultimate responsibility for the health of his insured patients and it is left chiefly to his discretion, to determine how much the consultant service should be used. A specialist to whom he sends an insured patient may not even send the latter on to another specialist without consulting the family doctor. The latter is charged, however, with the duty of scrupulous care in referring cases for hospitalization and specialist treatment, and he can be penalized[7] for careless, unnecessary, or "lazy" requests. The doctor is required in sending his patient to a specialist to state whether he refers him merely for diagnosis or for diagnosis and treatment as well. If the patient wishes specialist service beyond that which his family doctor believes is required, he may purchase it at his own expense at rates agreed upon in the collective contract between the Medical Society and the Federation of Societies in Copenhagen.

As in England, the health insurance doctor must provide medicaments, bandages, etc. which he needs in treating insured patients. These he requisitions on special prescription blanks, furnished by the Federation of Societies, which are honored by the firms with which the federation has made agreements.

[7] Apparently such disciplinary penalties have not been inflicted! The Medical Association Committee, however, keeps close watch on this as on other insurance situations. See p. 156.

In the metropolitan area approximately 415,000 insured men and women and their children[8] receive their general practitioner care from 252 health insurance doctors. The average insurance list of the Copenhagen "family doctor" is 1,616—in contrast to Britain's average of 975. Unquestionably the Copenhagen average would be lower were the medical panel open to all doctors instead of being a restricted list.

In making any comparisons with the situation of the British health insurance doctor, however, it must be remembered that in Britain the insured worker's wife, unless she is herself a wage earner, is not insured. In most cases, however, she is a potential patient of the worker's doctor just as completely as if she were on his medical list. In Denmark, on the other hand, the wives of the workers, whether wage earners or not, are supposed to insure themselves—which they customarily do.

The British doctor is permitted a medical list of 2,500, and the very popular doctor with a practice confined to wage-earning families often has this maximum. To this number must be added both the non-wage-earning wives and also the children in computing the doctor's "real" medical list in the sense of persons who normally look to him for medical care.

Certain of the popular general practitioners in the Copenhagen area (30 out of the total list of 252) have insurance lists that run from 3,000 to 4,000.[9] These lists already include most married women, who, as stated, are insured in their own right. A comparable "real" list, therefore, would add merely the children under fifteen, who, while entitled to insurance medical care, are not counted on the doctors' lists.[10] The difference between the situations of the two sets of doctors, so far as it relates to the number of their potential patients, is, therefore, not so great as would appear.

[8] Between 175,000 and 200,000 in number (estimated by the Directorate of Health Insurance Societies).

[9] See p. 135, third para. for comments on such practice.

[10] See pp. 136*ff*. on remuneration.

From a professional standpoint, however, the difference between the health insurance situation of a "family" doctor in Copenhagen and that of his fellow practitioner in a British city is marked. The British doctor must "go it alone," relying for consultant help for his insurance charges either on charitable assistance or on purchased specialist care which he knows is beyond the patient's means.[11] The Copenhagen general practitioner in contrast is equipped by the insurance scheme with a corps of well trained specialists in every field, to whom he can turn as the case demands. Similarly, through health insurance arrangements he can hospitalize his patients when they need it. In short, he can give his insured patients medical service that is as complete as contemporary medical science can make it.

Evidently, such a professional life stimulates the doctor to extend his interest beyond the purely curative problems of his daily practice. The journals of the Copenhagen Medical Association have been carrying an increasing number of articles on preventive medicine. These have been emanating not from public health officials or university professors, but from practising health insurance doctors who urge that "the times demand of the family doctor that he take his proper place as the unit in a great preventive drive for eliminating illness itself."

Remuneration in Copenhagen

In Copenhagen, as has been stated, the general practitioners and the specialists in the three fields of (1) eye, (2) ear, nose, and throat, and (3) "massage," are "regularly employed" doctors. Of these regularly employed doctors, the six "massage" specialists attached to the Federation's clinics are paid salaries. All the rest receive most of their remunera-

[11] The Insurance Acts Committee's advocacy that specialists' services be made an insurance benefit indicates that the British insurance practitioner is pushing toward an improvement of his medical position.

tion on the same basis as that in effect in Britain, i.e. capitation.

Unlike the British situation, however, the Copenhagen insurance doctors receive additional special payments. The family doctor is paid fees in supplement to his capitation income, for calls made on Sundays and holidays,[12] and the "regularly employed" specialists receive fees for *all* house calls. In other words, the capitation payment is considered the *general* practitioner's full remuneration for both office consultation and house calls, except those made on Sundays and holidays. The *specialists'* capitation payment, in contrast, is intended to compensate him merely for office consultations.

Payment for a house visit which calls for special fee is on the basis of one call only, even though more than one person is treated, when the doctor attends only the husband, wife, and their children under fifteen. Should another insured member of the household also need attention, however, this is paid for as a separate call.

In contrast to the regularly employed doctor who is paid a capitation sum for each member on his medical list (whether he sees him often or not at all), the "occasional" service specialists (i.e. specialists in fields other than eye, ear, nose,[13] and throat, and "massage") are paid on a fee basis for items of actual service rendered. Their fees come from a central fund maintained by the Federation of Societies. Each organization contributes 50 öre (a half-krone) per member per year to this fund, and the requests of a family doctor for services paid from this fund are supposed not to cost the fund more

[12] As stated previously, he does not have to make night calls, but may let the doctor employed for that purpose make his night calls. If he prefers to make his own night calls, he is not specially paid for these calls.

[13] Ear, nose, and throat service rendered in the home of the patient may be obtained, according to the 1936 agreement, not only from regularly employed physicians but also from physicians covered in the "occasional service" contract should the Federation make arrangements with such physicians.

than 50 öre per member on his list. Unless exceptional circumstances are shown, doctors who exceed this average may be asked to reimburse the "specialists' fund" to the extent of the excess. This rule is based upon the belief, apparently concurred in by the Medical Association, that such excess usually is caused by careless or lazy referring. Physicians with small lists (not more than 250) are not held to the half-krone rule but are required instead to obtain permission from the Federation in making each reference. Permission ordinarily is not refused unless the central fund has been unduly depleted.

The capitation figures and the special fees paid are fixed in the collective contracts between the Federation of Societies and the Copenhagen Medical Association. Deputies and substitutes are bound by these agreements equally with the men they replace. The remuneration scheme is an involved one. The scheduled figures are stated as of the year 1921 and then two-fifths of each figure set is adjusted with reference to the cost-of-living index number. The 1921 level of this number, which is used as a basis for adjustments, was the peak level of this number in the post-war period. For every 15 per cent which the index number moves above or below this basis, there is a 10 per cent increase or decrease in the adjustable portion of the remuneration schedule, i.e. in two-fifths of the remuneration sums agreed upon in the contract. It is stipulated, however, that under no circumstances shall the annual capitation of the family doctor fall below 6 kroner ($1.62) for each person on his list.

This complicated arrangement has been determined upon with the idea of preserving a balance between the doctor's need for a certain definite income which he can count upon from year to year (three-fifths remains fixed) and the justification of some variation in his income with substantial change in the cost of living. It is significant that the three-fifths which remains fixed is three-fifths of a figure deemed satisfactory in a period of peak prices, i.e. a figure set for the 1921 cost of living.

The cost-of-living index number is announced twice a year and if changes in the medical payment rates are called for, they go into effect for the next quarter.

Insured persons are divided into three groups for capitation purposes: (1) the ordinary insured, (2) insured "admitted with annotation of chronic disease," i.e. who had chronic disabilities when admitted to insurance, (3) insured who became active members of a society after passing their fortieth year. It is recognized that medical care of insured persons falling into the second and third classes may be expected to involve definitely more work on the average than in the case of the ordinary insured persons in Class I. Accordingly, there are three sets of capitation figures, and the doctor is paid 50 per cent more per year for responsibility for the care of each member of Class II, and roughly 35 per cent more for members of Class III than he receives for each ordinary insured person on his list (Class I).

Usually the doctor receives no special payment for treating children of the insured. The capitation amount has been agreed upon with this in mind. It is set at a figure considered high for the person without dependents and low for those with dependents, and is felt to average out reasonably. In cases where children are away from home, however, the full quarterly capitation sum is paid for each child to whom any medical service is rendered in the quarter. To receive this payment, the doctor must furnish details of such cases, with the name, address, and membership number of the person through whose insurance the children are entitled to medical benefit.

The annual capitation amounts for the *ordinary* insured are shown on the following table:[14]

[14] Schedules for Class II and Class III call for increased amounts as explained above.

Class I

(Ordinary insured persons, admitted to active membership before the age of 40 and without "annotation of chronic disease")

Type of Physician	*Annual Capitation*	*"Full list" Capitation Income*
General practitioners usual "full list" 2,000	9.00 kr. ($2.43)	18,000 kr. ($4,860)
Eye specialists usual "full list" 25,000	0.64 kr. ($0.173)	16,000 kr. ($4,320)
Ear, nose, and throat specialists usual "full list" 20,000	0.68 kr. ($0.184)	13,600 kr. ($3,672)
"Massage" specialists usual "full list" 30,000	0.68 kr. ($0.184)	20,400 kr. ($5,508)

It should be added that, as previously stated, the "full list" figure is not a hard and fast maximum. Thirty of the general practitioners in Copenhagen had lists of more than 3,000 in 1935, and many of the specialists received capitations for five to ten thousand more than the usual "full list" of their category.

Special fees are paid in addition to the capitation sums for calls which the family doctor makes on Sundays and holidays (covering a twenty-four hour period of 8:00 a.m. of the holiday to 8:00 a.m. of the day after). The fees are graduated according to the time of the calls, as follows:

Supplementary Fees

Sunday and Holiday House Calls, General Practitioner

8:00 a.m. to 8:00 p.m.	6.00 kr. ($1.62)
8:00 p.m. to 12:00 m.	8.00 kr. ($2.16)
12:00 m. to 8:00 a.m.	12.00 kr. ($3.24)

As in Britain, the doctor is paid every quarter. Every person on his list at the end of the quarter is counted in computing his check, which must be sent to the doctor within the first month following the quarter paid for.

The eye, ear, nose, and throat specialists' *special fees* (in addition to capitation) include a house call fee which is two and a half times the general practitioner's daytime rate if no surgical work is involved, and more than four times that rate when service of a surgical nature is required.

The specialist who is asked to give "massage" treatment in the insured patient's home receives considerably lower additional fees. Such home treatment is given only with special permission of the society acting on application signed by the specialist and stating in detail the extent of the contemplated treatment. Physiotherapy may be given either by the doctor personally or under his supervision by a physiotherapist authorized by the Danish Medical Association.

For obstetrical aid rendered by the family doctor either at childbirth or on abortion, he receives a special extra payment of 30 kroner. No distinction is made between instrument and non-instrument cases. An obstetrical *specialist* called in by the family doctor receives 20 kroner for a consultation and, for a delivery or assistance at the delivery of the child, he is paid 50 kroner or merely three-fifths more than the general practitioner's special fee. As in the case of other occasional specialist service, the family doctor is held responsible to call such assistance only when circumstances justify.

The obstetrical service fees paid to the family doctor and to the specialist are provided from a special maternity benefit fund.

Deputies who substitute during the absence of a general practitioner receive for all house calls the same special fees as are paid to the usual practitioner for calls made on Sundays and holidays, i.e.:

6 kroner	($1.62)	between	8:00 a.m. and 8:00 p.m.
8 "	($2.16)	"	8:00 p.m. and 12:00 p.m.
12 "	($3.24)	"	12:00 p.m. and 8:00 a.m.

Young doctors, when waiting to be put on the insurance list and when first on the list, while they are building up an insurance practice, often apply for deputy work, i.e. "night watch"[15] and Sunday and holiday assignments. If a doctor substitutes for several physicians or is himself an insurance practitioner, he may not without special permission from the

[15] The doctor on "night watch" gets paid for visits even to his own patients, when made in his substitute capacity on Sundays, holidays, and at night.

Federation of Societies undertake responsibility for substitute practice involving lists of more than 4,000 insured persons.

Fees for miscellaneous services,[16] which are considered outside insurance obligations and which, therefore, may be exacted directly from the patient are regulated in amount by the collective contracts. The fees for such services add in a small way to the health insurance doctor's income.

Many of the health insurance doctors have a "private practice" in supplement to their insurance work. This is apparently the case with most of the specialists and with those general practitioners who have lists of less than 2,000. A few of the latter who have between 2,000 and 3,000 on their insurance lists also do some private work. Health insurance officials believe that those with lists of 3,000 or more do not have any practice other than health insurance practice.

Collective Contract II

The "occasional service" physicians whose fees are regulated, with other conditions of their health insurance service, by Collective Contract II[17] receive fees that range, if the work is office service, from 8 kroner for a first consultation to 10 kroner for single treatments, i.e. approximately the annual capitation fee paid the family doctor. The fee is doubled for complete examination and written report of diagnosis[18] and 50 per cent more is added to this in case of minor operations. Each of these fees is increased 50 per cent when service is rendered in the patient's home. X-ray examinations are paid at rates from 15 to 40 kroner ($4.05 to $10.80).

[16] These include giving (1) general health certificates at 3 kr.; (2) medical examination for admission into a health insurance society at 3 kr.; and (3) certification for entry into an institution for abnormals at 12 kr.

[17] The present contract became effective July 1, 1936, to be effective until 1941 and thereafter until revoked by six months' notice from either the Federation of Societies or the Medical Association.

[18] cf. the "1 guinea" ($5.25) specialist consultant service in London, p. 86 *supra*.

The specialists who serve on the hospital staffs, where all serious operations are performed, of course add to the physicians who render services to the insured families. These doctors are paid only by the hospitals and not directly by the Insurance Societies or their members. As nearly 90 per cent of the population received insurance benefits and all serious surgical cases and most of the serious medical cases are hospitalized, the better part of the time of the hospital staff is spent in operating upon and treating insurance patients.[19] The salaries of the medical staffs, therefore, should be mentioned in discussing the remuneration of the health insurance doctors.

In 1935 the forty-six Health Insurance Societies in the Copenhagen area paid their 252 family doctors 2,874,720 kroner ($776,174.40), an average (arithmetical mean) of 11,408 kroner ($3,080) per doctor. This amount constituted 2.7 times the full-time annual earnings of the skilled worker in the metropolitan area and four times the similar earnings of the unskilled wage earner. The *median* income was 10,934 kroner ($2,952). The largest capitation income, i.e. not counting "extra fees," which any one *general practitioner* received from health insurance practice was 28.994 kroner ($7.828). This was over 7 times as much as the skilled worker could earn in a year and over ten times what could be earned by the unskilled man. As the Copenhagen general practitioner has his office in his home, and does not in most cases employ a nurse, most of his income is "net."

It is interesting to compare this situation with that of American general practitioners as revealed by the Hoover "Committee on Costs of Medical Care."[20] During the peak

[19] Not only in Copenhagen but elsewhere in Denmark, rural physicians perform operations in communal hospitals on invitation. See p. 146 *infra*. Some of the time of the hospital doctors is, of course, given to non-insured, although most of it is spent in treatment of the insured and their families. See p. 117 *supra*. Statistics of these salaries are not available in Denmark. Only total figures of the operating costs of the hospitals can be given. See p. 118 *supra*.

[20] See Committee on Costs of Medical Care. Pub. No. 24, *The Incomes of Physicians*, pp. 109, 117.

year of "prosperity"—1929—the average (arithmetical mean) net income was $3,900 and the median net income was $2,900. This refers to *all* American general practitioners—those who attend the well-to-do as well as those who attend working families. This average was just about 25 per cent more than the maximum annual earnings of the highest paid skilled mechanics in an American city in 1935, while the median figure was actually less than the earning capacity of such a worker.

As the distribution of medical income from health insurance is determined chiefly by the size of the medical lists of the doctors, the distribution in 1935 of the insured among the Copenhagen doctors is of interest:

DISTRIBUTION OF INSURED AMONG DOCTORS
GENERAL PRACTITIONERS. COPENHAGEN, 1935

Number of insured on medical list	Under 500	501-1,000	1,001-1,500	1,501-2,000	2,001-2,500	2,501-3,000	3,001-3,500	3,501-4,000	Over 4,000
Number of doctors	43	38	43	38	39	21	19	10	1

Translating this table into one showing the corresponding medical incomes[21] from health insurance practice produces the table below:

DISTRIBUTION OF HEALTH INSURANCE INCOMES
GENERAL PRACTITIONERS

Size of Income	*Number of Doctors*	*Per Cent of Doctors*
Under 3,530 kr. ($953)	43	17.00
3,530-7,060 kr. ($953-$1,906)	38	15.00
7,060-10,590 kr. ($1,906-$2,859)	43	17.00
10,590-14,120 kr. ($2,859-$3,812)	38	15.00
14,120-17,650 kr. ($3,812-$4,765)	39	15.04
17,650-21,180 kr. ($4,765-$5,718)	21	8.33
21,180-24,710 kr. ($5,718-$6,671)	19	7.53
24,710-28,240 kr. ($6,671-$7,624)	10	4.00
Over 24,240 kr. ($7,624)	1	0.50

In addition to the amount paid for family doctor service in Copenhagen, the sum of 935,809 kroner ($252,668.43) was

[21] Assuming "extra" fees are evenly distributed, computed at 7.06 kroner per person on list—converting the krone at 27 cents, actual capitation in 1935 was 6.35 kroner. For further details of Copenhagen health insurance incomes, see De Samvirkende Sygekasser, *Köbenhavn Beretning for Aaret*, 1935.

paid to specialists, some of whom were paid more than 30,000 kroner ($8,100) for this part of their practice. The above figures, of course, do not include the remuneration of the hospital staff doctors who attended the insured patients. These patients accounted for more than 1,600,000 hospital days (the very high rate of nearly two and two-thirds days per insured beneficiary) in 1935, a condition which, as previously explained, is related to the deliberate public policy[22] of institutionalizing sickness cases when congested home conditions make home nursing difficult.

[22] Dr. Frandzen, Denmark's Director of the National Health Service, called attention of the author to the fact that the high hospitalization rate is a direct reflection of this policy. The wife, he stated, can care for the husband who is ill when there are no children. When there are children or when the wife is bedridden and needs constant attention, crowded apartment house accommodations make home nursing impracticable and hospitalization, not otherwise required, is encouraged as a public health policy. See p. 124, n. 4 *supra*.

XV

THE HEALTH INSURANCE DOCTOR IN NON-METROPOLITAN DENMARK

In General

AS previously stated, outside the Copenhagen area any general practitioner belonging to the Medical Association who wishes to practise on the terms set by the regional collective agreement, may do so. The terms of service of such specialists' attention as may be offered are fixed, for the most part, not by regional collective agreements, but in one contract between the Central Union of Federated Societies and the central committee of the Danish Medical Association. Each health insurance society determines individually whether to provide such services and must, of course, abide by the terms of the central contract in so doing. About 750 funds outside the metropolitan area have in fact made contracts with ophthamologists and otologists.[1] Although no central record is kept of the extent to which these societies actually permit their insured members access to these specialists, it is believed at the Directorate of Health Insurance Societies that the urban societies at least provide such special help freely as needed.

The regional agreements governing family doctor service differ in certain details, but their major provisions are similar. The practitioner is obligated to render all necessary "family doctor" service, both at his office and at the patient's home. Unlike the Copenhagen family doctor, the rural prac-

[1] Although orthopedic surgeons, dermatologists, and "massage" specialists are arranged for in a few of the small cities, the only services generally regulated by the central contract are those of eye, ear, nose, and throat doctors and "massage" specialists.

titioner must do work that requires the assistance of a colleague and he must assist a colleague on demand. He must extract teeth when no insurance arrangement with dentists exist. Many, but not all, of the non-metropolitan communal hospitals have a surgeon for medical director. The country doctor is expected to perform major operations when necessary, and render many other services which the metropolitan health insurance doctor is free to turn over to a specialist or to a hospital staff.[2]

The agreements provide for two alternative methods of payment of practitioners, Tariff I, the Copenhagen method, i.e. chiefly capitation with supplementary fees for special services, and Tariff II, payment for each unit of service rendered according to a fee schedule.

Ordinarily the selection between Tariff I and Tariff II is made by a local Federation of Societies rather than by an individual health insurance organization. Any society, however, may elect to use either tariff (or to change from one tariff to the other) even when the general area has not so decided, unless there are special conditions involved in medical service to its members which would make such an arrangement unreasonable. The physicians concerned are permitted to challenge as unreasonable a single society's selection of method of payment. The local Federation of Societies then takes up the question with the local Medical Association. If they are unable to come to an agreement, the question is referred to Denmark's Central Council of Arbitration[3] for decision.

Similarly, if the doctors practising for a society desire to change from one tariff to the other they may submit the request to their local Medical Association. If it approves the change, the association then attempts to negotiate the change with the local Federation of Societies, and negotia-

[2] The central hospitals have staffs of specialists, including of course surgeons. See p. 158 *infra* for discussion of these hospitals.

[3] See p. 116, n. 11, for composition, etc., of this arbitration council.

tions failing, sends the question to the Central Council of Arbitration for determination.

No newly organized society that admits members without medical examination is allowed to select capitation without obtaining permission from the local Medical Association.

Tariff I (capitation) is used by all the societies in the urban areas and by 350 societies in rural areas which include about 900,000 members, i.e. more than half of the insured persons outside of Copenhagen. Tariff II is used by 1,100 rural organizations with a membership of 800,000. As there are some 2,100,000 insured persons in Denmark, it follows that the doctors serving 38 per cent of the insured are paid on a fee basis, and the practitioners who attend 62 per cent of the insured receive capitation payments. There is, moreover, a definite trend toward the adoption of the capitation method. Each year a larger number of societies have employed this system of paying their health insurance physicians.

In 1935 the general practitioners outside Copenhagen were paid 14,887,648 kroner ($4,019,674.96) for health insurance service. Assuming that every general practitioner in practice in these areas (there were 1,124) served some insurance patients, this meant an average annual income from health insurance work of 13,245 kroner ($3,576.15) per doctor. This is three and three-fifths times the maximum earnings of the skilled worker of the towns and more than four times the income of the full-time skilled worker in the rural areas.[4] This estimated average is the lowest possible figure, and of course if the number of doctors participating in health insurance practice was actually less, the average income was greater. No official bureau has actual record of the exact number of general practitioners in health insurance practice outside of Copenhagen, but it is believed by the government's

[4] Such a town worker's maximum income is estimated at 3,600 kroner by the government statistical bureau, the rural workers at 2,800 kroner.

supervisory office that "practically all" such doctors have some insured patients.[5]

In addition to health insurance income from the societies, the communes paid the travelling expenses incurred by these doctors in attending to their health insurance practice.

The collective contract for Jutland governs the health insurance practice of the doctors who serve the most populous district outside of Copenhagen,[6] containing more than half of the insured who live elsewhere than in the capital. It is, therefore, in itself the most important of the agreements other than that which functions in the metropolitan area. Further, as previously remarked, it differs from the other five contracts in only the less significant details. Its terms in consequence may be deemed illustrative of all the collective agreements operative in Denmark outside of her capital.

In Jutland

The doctor who puts his name on the panel for health insurance work, normally is available to patients within a ten-kilometer radius *by land*.[7] He may, however, by giving notice through an appropriate committee of his Medical Association, confine his practice to the town in which he lives, or to a specified rural area. Further, if a doctor has less than ten members of a given society on his medical list, he is not obligated to practise for that society unless there is no alternative panel doctor accessible for these members.

If a society has insured members who do not live within ten kilometers of any health insurance doctor, the nearest panel doctor within a distance of fifteen kilometers is obligated to serve.

The extent of choice of doctor permitted to the insured person may depend upon the method of payment of physician which is used by the society.

[5] Information furnished by Direktor Borberg in Copenhagen, August 1936, at Directorate of the Health Insurance Funds.

[6] One-fifth of Denmark's insured live in Copenhagen.

[7] Because of Denmark's many islands, this is specified. (This is 6¼ miles.)

In all cases where the capitation basis is used, members are required, as in Copenhagen, to select a doctor for a year at a time (January to January) and a married couple must have the same doctor. Ordinarily they may choose any doctor within ten kilometers.

When the doctors are paid on a fee basis, usually the insured may go freely from one doctor to another within the permitted radius. For economy purposes, however, societies are permitted to require choice for a full year even with the fee basis. Experience, they believe, has demonstrated that the patient will make fewer demands for medical service when he is held down to one physician than when he (mostly *she*, it is alleged) is permitted frequent change of medical scene. The list of doctors available for choice is posted at the society's office and many of the organizations also furnish annual printed slips of such information for their members.

Ordinarily the doctor, at his own expense, must provide a deputy for periods when he is to be away. An exception is made, however, for short absences (not more than three days) in localities with only one resident physician. Should an insured person need medical attention under such circumstances, he may call in another health insurance doctor, and the society must pay him according to Tariff II.

In case the doctor selected dies or gives up his practice, insured persons on his list may leave the substitute or successor whom he designates, at the beginning of the following quarter. Similarly, when a deputy is employed for longer than six weeks, for reasons other than illness or post-graduate medical study, the members of his list may choose another physician at the first of the next quarter.

In case the doctor practising under Tariff I is required to treat more than one person when making a specially paid for house call, he is paid for only one such call. If, however, he is practising under Tariff II (i.e. fees), he is paid additionally on an office consultation basis for each extra person attended.

As in Copenhagen, except in emergencies, the insured in

applying for treatment must identify himself either by presenting his membership book or by some alternative method approved by his society, such as presentation of a medical report card. The doctor may demand a fee if identification is not made. The first medical service to an eligible insured person, however, will be paid for by his society even if the insured carelessly forgets his credentials.

The patient is required, as in Copenhagen, to make all possible concessions to the doctor's convenience. Unless the physical condition of the patient prevents, the doctor is to be seen at his office during the hours set for consultation. Requests for house calls are to be in before 10:00 a.m. in the country and before 9:00 in town practice, except when emergencies arise. Further, if the insured calls for medical service at night or on Sundays and holidays in other than urgent circumstances (i.e. accident, sudden illness, or relapse) he must pay for it himself. The doctor, however, even in these cases, gets a certain amount of protection by the guaranty from the society of 50 per cent of such bills in case they remain unpaid for two months after presentation. The other 50 per cent he must collect from his patient as he does his "private practice" bills.

The patient must, as in Copenhagen, obey medical instructions and accept hospitalization when the doctor orders it. He is specifically obligated to undergo the free medical treatment which is furnished by the commune in all parts of Denmark in case of communicable disease (including venereal disease). The patient's refusal to follow orders, however, relieves the Jutland doctor only of the duty of visiting the patient in his home and not of *all* medical obligations as under the Copenhagen arrangements.

Remuneration (Jutland Contract)

As in Copenhagen, remuneration schedules are stated in the terms of the 1921 cost of living and adjusted in the same way. Forty per cent of all stipulated remuneration is subject to a 10 per cent increase or decrease for each fifteen points that the index number is above or below the 1921 level of 264.

The capitation amounts of Tariff I for general practitioners are identical with those effective in Copenhagen.[8] More services are specially paid for, however, than in the capital, and the amounts paid are frequently larger. Not only is obstetrical aid (including abortion cases) on a fee basis as in Copenhagen, but also: (1) First treatment of serious injuries or burns; (2) setting and bandaging major fractures and dislocations; (3) assisting another health insurance doctor; (4) emergency house calls required to be made immediately or during scheduled consultation hours; (5) extraction of teeth; and (6) night calls.

Charges similar but often larger than those allowed in Copenhagen are permitted for the issue of certificates needed in entering mental hospitals, tuberculosis sanatoria, and certain convalescent institutions, for vaccination and death certificates, and for medical certificates needed by applicants for admission into the Health Insurance Societies.

The eye, ear, nose, and throat specialists, if paid by capitation, receive a little more per insured person on their lists than in the metropolitan area with certain "extras" as in Copenhagen.

The fee schedule to be used when Tariff II[9] is employed as the basis of the general practitioner's remuneration is elaborate, covering services ranging from a consultation by telephone at 2.50 kroner to a tonsil-and-adenoid operation or first treatment of serious injuries at 20 kroner. The latter is about two days' wages of the provincial metal worker, more than three times the daily wage of the provincial engineering laborer, and four times that of the agricultural

[8] i.e. 9 kroner per person on the doctor's medical list (18 kroner for a married couple) with 50 per cent additional for persons who entered with a chronic disability and 35 per cent extra in the case of persons who became active participating members after reaching the age of forty. See pp. 132-3 *supra*.

[9] A higher fee schedule is used for the same services in case of specialists.

laborer.[10] Fees for house calls increase in amount with the inconvenience of the hour.

Rates paid by the funds to the specialists under Tariff II, in contrast to the arrangements in Copenhagen, where hospitals with full specialist staffs are available, include fees for operations performed as "guest surgeons"[11] in both hospitals and hospital clinics, as well as rates for operations performed at the doctor's office or in the patient's home. The operating fees range from 13.5 kroner, which is three days' earnings of a Danish agricultural laborer, to 135 kroner, or the equivalent of more than a month's pay for a farm laborer.[12] The rates for office and house calls are lower than in Copenhagen.

When capitation (Tariff I) is used, the doctors must be paid, as in Copenhagen, every quarter, unless arranged otherwise locally with agreement of the local Medical Association. The list of the quarter is taken as of the sixteenth day of the second month of the quarter.

Payment under Tariff II (fee basis), and also the fees for extra services paid in supplement to Tariff I are due quarterly, semi-annually, or even annually, as arranged with the individual physician. Dilatory settlement of accounts is penalized. The doctor must render account within five weeks of the period it covers or suffer a 10 per cent deduction. In like fashion, if he receives his check more than a month after his statement of services rendered has reached the society, the physician is entitled to an additional 10 per cent. In case certain items are challenged, only the disputed portion of the account may be withheld.

[10] In 1934 and 1935 the provincial forge worker earned 11 kroner, the engineering laborer 6 kroner, and the agricultural worker averaged 4 kroner 60 öre. Denmark, *Statistisk Aarbog*, 1936, pp. 124-6.

[11] With permission of the communal hospital doctor.

[12] These are the wages of a farm worker who boards himself. When board is provided, wages are reduced one-third.

XVI

PAPER WORK INVOLVED IN DANISH HEALTH INSURANCE PRACTICE

As in Britain, the Danish health insurance doctor has "certification" responsibilities. His statement of the insured person's incapacity for work is accepted, as the British doctor's ordinarily is, in establishing the right to cash sickness benefit. The responsibility of the Danish doctor, furthermore, is more complete than that of the health insurance doctor in England, as there is no medical referee system in Denmark. Weekly reports for the society must be signed by the doctor while the worker is at home sick. Similarly, the doctor must issue a "signing off" report when he deems the insured once more ready for work.

In contrast to the British situation,[1] neither the Copenhagen agreement nor the law specifies whether diagnosis shall be included in the incapacity certificates, and practice is not uniform. The disability is often specified in Latin and frequently abbreviations of the Latin terms are employed. All diagnostic information is, of course, ordinarily confidential, and administrative officers are required under pain of severe penalties to observe absolute secrecy with reference to the medical reports received by the societies.

In case of tuberculosis or venereal disease, the attending physician, whether in insurance work or in private practice, is obligated to notify the appropriate health authorities. Also, he is required, as are teachers, societies, etc., in the event of discovering *any* disability tending toward invalidity, to report the case to the Invalidity Institute, the governmental agency which administers Denmark's compulsory Invalidity

[1] See pp. 49-50 *supra*.

Insurance Law. The Institute is furnished this information to enable it to prevent invalidity when possible, by providing special medical treatment or other necessary care. The Institute pays for this treatment, when there is a reasonable chance that it may prevent invalidity or lessen its degree.[2]

If the case has been turned over completely to a specialist, he and not the "family doctor" must do the certifying. If, however, the specialist is acting only in a consultant capacity, the general practitioner is responsible for the paper work.

If an insured person, for any reason, is arranging for treatment from a non-health insurance doctor (for which he must pay individually) the regularly employed family doctor in the subdistrict must issue certificates only on special request from the society involved. This situation is an exceptional one.

When the attending doctor, whether general practitioner or specialist, wishes to hospitalize his patient, he must fill out a prescribed form, which includes a diagnosis of the case.

Issuance of weekly reports to enable the insured to obtain "continued" sickness benefit due from the commune, and the signing of a miscellaneous group of certificates, such as disinfection certificates, school certificates to exempt the child from certain types of school work, death certificates for use in collecting funeral benefits from insurance societies, etc., are included in the doctor's health insurance duties.[3]

[2] Otherwise, it may not pay for treatment. So long as the person can make a fair contribution toward his self-support, he gets treatment through health insurance. When he is in receipt of an invalidity pension, medical care for which the Institute cannot pay must be provided by the communal authorities. It is in dispute whether the latter may exact a means test before providing medical care. Dr. Hulck, director of the Invalidity Institute, construes Sec. 69 of the Social Insurance Invalidity provisions as requiring the communal authorities to furnish the care as a right of the pensioner and without a means test, but this is resisted by certain communes. No test case has yet been brought.

[3] In addition, as stated previously (p. 135 *supra*), the collective contracts stipulate fees which may be charged for other reports and paper work which the physician may be called upon to arrange for an insured person, such as a certificate to obtain "sick pay" from an employer who makes such provision for his workers. Certificates other than those for which rates are set must be furnished by the doctor at the minimum rate set for such work in the fee schedule of the Medical Association.

In Copenhagen, "control slips" must be signed when the doctor makes a call that involves a "control" payment by the member, and an itemized statement of all work specially paid for must be presented in all but physiotherapy cases within a week of service rendered. However, physiotherapy treatment which is given in the home of the insured may be reported quarterly.

The paper work outside the metropolitan area varies with the method of paying the physicians. If capitation (Tariff I) is used, the doctors are, with minor exceptions, in the same position as the Copenhagen practitioners.[4] If the fee system (Tariff II) is employed, the doctor must, of course, keep many more records. Not only disability certificates and control slips, but also detailed records of all units of medical service rendered must be prepared and forwarded to the societies, as remuneration is adjusted in accordance with the service rendered. The collective contracts specifically obligate the doctor, when paid on a fee basis, to indicate the time, date, etc., and also, in *Danish*, not Latin, the nature of the service rendered.

As in Britain, standardized forms for all reports, certificates, control slips, etc., are furnished the physicians for use in health insurance work; thereby reducing the insurance paper work to the minimum.

[4] e.g. in the non-metropolitan area giving a death certificate is not an insurance obligation.

XVII

DISCIPLINARY CONTROL OF THE PHYSICIAN

IN contrast to British arrangements, there are no medical officers to act as referees and to give medical supervision to the clinical records kept by the Danish health insurance doctors. It is not denied, of course, that an occasional Danish doctor is less scrupulous or less careful than he should be about the financial interests of a Health Insurance Society nor is it even asserted that under no circumstances does a Danish doctor fail to fulfil his strictly professional obligations to his insured patients. It is, however, the conviction, not only of the Danish doctors but of everyone connected with health insurance administration, that matters which cannot be adjusted by an informal telephone call or conference are not apt to occur; that the machinery set up to hear and adjust the occasional complaints against the health insurance doctors is adequate to meet all needs; that to set up a scheme of supervisory medical officers would involve sheer waste of time and money.

The record bears out these contentions. In more than twenty years not one Danish doctor has been denied the right to continue health insurance practice. Nor has any Copenhagen insurance practitioner in this period, been penalized officially by being *required* to forego certain of his remuneration or to reimburse the society for "unnecessary costs," as the disciplinary machinery permits. (This does not mean that adjustments have never been made between the doctor and the society, in formal or informal conference.)

Disciplinary Machinery in Copenhagen

It will be remembered that the Copenhagen Collective Contract on health insurance medical service calls for an

Appointment or "Vacancies" Committee,[1] composed of three representatives of the Copenhagen Medical Association and three of the Federated Health Insurance Societies. This committee includes among its powers and duties that of acting on complaints from a society that a doctor has imposed upon it unnecessary expenses. Such complaints may rest upon claims that the physician has been guilty of extravagant prescription, or has made unnecessary calls requiring special payment, unnecessary references to specialists, or unjustified hospitalization of a patient. After investigation, which in practice would mean special consideration by the members of the committee who represent the organized profession, the committee may, as in Britain, "surcharge" the doctor in an appropriate amount by deduction from his next quarterly check. The doctor may appeal against such a surcharge and the society may appeal if the committee itself fails to surcharge. The matter then would go to the Arbitration Committee described in the next paragraph.

If other formal charges should be brought against a doctor in Copenhagen, either by a member or by a society's agent, the rules require that they be filed in formal written form with the administrative officers of the society involved, within six weeks of the episode complained of. These officers, unless convinced that the charge is frivolous, must present the complaint to the doctor within a month of receiving it. The doctor, in turn, is then entitled to two weeks to file a statement on his own behalf. The officers then must endeavor to settle the question by direct negotiation with the doctor. That failing, either the physician or the society may submit written statements to the Medical Association and the Federation of Health Insurance Societies, each of which will appoint a mediator. If the mediator cannot adjust the matter, it must go back to the two organizations (the Medical Association and the Federation of Societies) for decision. Finally, should there be a case of deadlock between these organiza-

[1] See p. 121 *supra*.

tions, the matter would pass to formal arbitration by a standing Arbitration Committee organized specially for the Copenhagen area and composed of two representatives (and two alternates) from each of the two organizations and a fifth member chosen by the four. If the four cannot agree upon this fifth appointment, they must select a chairman, who appoints one of the Danish judiciary as the additional member. Rules require this fifth "independent" arbitrator to be unconnected with either organization. He serves for a four-year renewable period.

The Arbitration Committee is empowered to make its own rules and is specifically authorized to negotiate a settlement between the parties at any stage in the proceeding. Decisions are binding and non-appealable. If necessary, the committee may call upon the courts to enforce compliance with its orders, and enforcement costs are assessable to the resisting party.

Disciplinary Machinery in Non-Metropolitan Denmark

In the non-metropolitan areas, the disciplinary procedure differs from that arranged in the capital in the elimination of the Vacancies Committee and in the substitution of a national Central Arbitration Board for the local Arbitration Committee of Copenhagen. If an insured person complains of his doctor, his society first hears the doctor's explanation. If convinced that the complaint needs investigation, the society then forwards the charges to the Central Council of the District Federation of Societies. Similarly, a society that feels itself imposed upon by a doctor sends its case to this same central council. The doctor who has a grievance files his statement with the Sick Benefit Committee of the District Medical Association.

The Federation's Central Council and the Medical Association's Sick Benefit Committee attempt to settle all such situations by negotiation. On failure of the two agencies to agree upon the disposal of a case, the matter goes for settlement to the Central Board of Arbitration set up by the 1933 Social Security Act.[2] It has the same powers as those of Copenhagen's Arbitration Committee, described above.

As in Britain, the Danish rules provide that a doctor against whom charges are filed may not escape his responsibilities to the disciplinary authorities by resignation.

No doctor in the non-metropolitan area, and in Copenhagen no health insurance doctor who has been employed for longer than a year and a half (during which period dismissal is permitted on three months' notice), may be removed from the panel except for cause. The doctor who believes the cancellation of his contract unjustified or believes that he has been otherwise injured by a society or by an insured person, files a complaint with his Medical Association and with the Federation of Societies. It is then adjusted by the steps outlined above for settlement of complaints against a doctor. The ultimate responsibility for directly depriving a health insurance doctor of the right to continue in insurance practice thus rests with the local Arbitration Committee in Copenhagen and with the Central Board of Arbitration elsewhere.

Disputes between the Medical Association and the Federated Societies over the interpretation of the terms of the collective contracts also go to the respective arbitration bodies for settlement.

In the twenty-year period between 1917 and 1936 the Central Arbitration Board was called upon to meet in every year but two (1919 and 1920). It held thirty-eight sessions in the eighteen years in which it convened and considered a total of seventy-two[3] cases. These sessions and cases were distributed as follows:

[2] See pp. 104-6 *supra* for discussion of this act.
[3] Transcript of case records furnished by Direktor Borberg.

Year	*Session*	*Cases*
1917	1	2
1918	1	1
1921	1	4
1922	3	6
1923	3	6
1924	1	1
1925	2	4
1926	2	4
1927	3	4
1928	3	6
1929	2	5
1930	2	3
1931	2	4
1932	2	5
1933	2	2
1934	2	3
1935	3	7
1936	3	5
TOTAL	38	72

Of the seventy-two cases adjudicated, fifty-two were disputes over the meaning of allegedly ambiguous provisions of the agreement, one involving the propriety of oral discussion of and publication of grievances against a doctor instead of pursuing the complaint procedure of the agreement. Fifteen were disputes over proposed change from Tariff II to I. The latter were resolved by permitting the change in one instance only in part of the geographical area concerned, by refusing it altogether in three, and permitting it altogether in eleven cases.

In five cases the doctor was accused of deliberately causing the society unnecessary expense, and his bill was challenged in consequence. In three of these cases, the specific ground of complaint was that the doctor kept the insured person at home and ran up a visitation bill when he should have hospitalized the patient (hospitalization would have cost the fund far less). In one instance the additional misconduct of a 25 per cent overcharge for mileage was alleged. Two of these hospitalization disputes were resolved against the doctor and one in his favor.

In two cases the doctor was accused of making wholly unnecessary visits. One of these disputes had been settled (through conciliation proceedings) by the fund paying and the doctor accepting a reduced amount in lieu of the bill presented. The doctor subsequently repudiated this compromise and, with the support of the local Medical Association, carried the case to the appellate arbitration body. He argued that the informal conciliation proceedings were void as violative of the collective medical agreement's requirements of written complaint and formal hearing in all cases in which the fund alleged medical impropriety. This contention was held to be an unsound one and the doctor was forced to abide by the compromise he had accepted. The other "unnecessary visitation" case was decided in favor of the physician.

In addition to these five cases, two of the "interpretation of agreement" cases involved a challenge of bills presented by the doctor, and implied at least a questioning of the doctor's good faith. In one case, the bills were questioned on the score that control slips were not issued or asked for by the doctor. In addition there seemed to be a question about the patient's need of the consultations. The doctor had not asked her to come to his office, and it was not established that her physical condition necessitated this medical attention. This case was resolved against the doctor. In the other case, the questioned bill concerned telephone consultations for which no proof was offered beyond the presented bill. This case was resolved in the doctor's favor.

In summary, it may be said that in the twenty-year period since the board's creation, not a single case involving such professional misconduct as neglect of a patient has come before the board. Only seven disputes which involved alleged overcharging or improper charging by the Danish health insurance doctor have reached the formal arbitration stage. In only four of these cases was the doctor actually held guilty. In two instances, he was granted his full claim. In the remaining case, he was held to the compromised reduced bill which he had agreed orally to accept in conciliation proceedings. This record definitely bears out the belief of the profession and of lay administrators that the Denmark health insurance scheme has no need of additional supervisory machinery.

As in Britain, at every step of procedure which may affect the health insurance doctor in Denmark, it is arranged that representatives of the organized medical profession shall play an important part. The uniformly high standard maintained in the medical profession in Denmark, trained as its members are in one excellent medical school, removes all likelihood of serious disciplinary situations involving the health insurance doctor. Moreover, the practitioner who tends to carelessness is subjected in practice to the control of the organized profession, which apparently "quietly takes care of" his case within the organization. As no physician may be employed in health insurance work unless he belongs to the Medical Association,[4] expulsion from the organization would terminate health insurance employment, wholly aside from the disciplinary procedure outlined. Through the control which this situation places in the hands of the Medical Association, quite independently of any machinery set up in the collective contracts or in the health insurance law, the organized Danish medical profession can and does keep the health insurance doctor up to an enviably high standard of responsibility in all phases of his professional obligations.

[4] Or has special permission from the association. See p. 119 *supra*.

XVIII

ATTITUDE OF THE MEDICAL PROFESSION

Attitude Toward Certification Responsibilities

THE profession in Denmark can hardly be said to have an "attitude" toward their certification responsibilities in the sense of regarding them as something apart from their ordinary professional obligations.

Even more so than in England, where the existence of a corps of medical referees at least emphasizes the certification duty, the Copenhagen doctor is puzzled by the suggestion that he might think his certification duty an imposition. It may be doubted that such an idea has ever occurred to him. He reasons, as does the British doctor, that his patient needs his cash benefit to pay for his food when he is ill and cannot work; only the attending doctor knows whether his patient is in a condition of capacity or incapacity; ergo, naturally, the doctor must certify this condition.

The Danish people are reputed to have a generous quota of moral responsibility in their make-up. The fact that neither the health insurance societies nor the health insurance physicians even feel the need of a medical referee corps indicates that the moral responsibility involved in the certification duty is not only shouldered but carried comfortably by the health insurance doctor.

Naturally, with such a happily adjusted situation, the Danish doctors regarded the question of whether state medical officers could satisfactorily perform the certification function as a purely academic one. As such, they dismissed it, without exception, as did their fellow practitioners in Britain, as wholly unworkable. The only difference, indeed, between the Danish response and the British response was that in Britain this conclusion was occasionally remarked with regret.

Why the Profession Approves Its Health Insurance Arrangements

The Danish medical profession, in expressing wholehearted satisfaction with health insurance, stresses the broad public health and social aspects of insurance practice and its happy effect upon the professional life of the doctors. They also admit, however, that its financial advantages to the doctors play a definite part in their approval of insurance practice, as it does in the comparable satisfaction of the British profession. Although Danish health insurance has been in operation for forty-five years, the doctors about fifteen years ago apparently had an object lesson in the superiority of medical incomes under health insurance, over private practice returns from comparable patients. As a result of a bitter controversy in Jutland over the scale of payment of medical health insurance practice on the expiration of the collective contract in this particular year, no renewal contract was sent to the Minister for approval. In consequence, the usual medical benefit was in abeyance for a year. Doctors were on "private practice" terms with the insured patients, who were entitled to reimbursement from their societies of from 50 to 75 per cent of the medical bills they paid.

In consequence, to quote the exact words used by the distinguished practitioner, Dr. Johan Kuhn, of Copenhagen: "During these twelve months it was noticeable that the demands for medical service diminished perceptibly; and despite the fact that all the doctors hardened their hearts, going even so far as to collect their bills through bill collectors, and despite the fact that also now they were free to name their own fees and that they no longer gave reduced rates to the insured, despite all this, this controversy (from which the doctors in other respects came out "on top") cost them about 1,500,000 kroner [$405,000]!"[1]

This meant an average loss that year of about 3,700 kroner—approximately $1,000 per health insurance doctor![2] At such a price the prerogatives of private practice might tend to lose their charm even were the general conditions of insurance practice deemed objectionable, which, of course, they were not in this case. On the contrary, the Danish practitioners find health insurance practice most attractive from the highest professional standpoint. They have reached this conclusion with even more justification than their professional brethren in Britain because of the generous terms of the Danish medical benefit. Equipped with insurance benefits, the doctor in Denmark can hospitalize his "unpropertied" patient and otherwise arrange his treatment largely as his professional judgment dictates, instead of always trimming it to what the patient can afford. The Danish doctors emphasize that this situation makes for a happy professional life. They also stress the point that the absence of money passing directly between insured patient and doctor brings improvement in the doctor-patient relationship. Indeed, these practitioners found it hard, as their British colleagues did, to believe that one could suggest in good faith that the worker might love his doctor the more for being a "perambulating doctor bill."

Apparently there has been a relatively recent conversion on the part of many Danish doctors to capitation as against the fee system of remuneration for insurance services. The

[1] This incident was described by Dr. Kuhn personally in an interview with the author. He referred also to his public address in Paris in which he had related the facts. See pp. 13*ff*. of an address delivered by Dr. Johan Kuhn, then Secretary of Danish Medical Society (now social and technical editor of the *Danish Medical Journal*) in Paris in 1927 at the occasion of organizing the International Bureau of Hygiene (League of Nations)—"La Pratique de l'Assurance—Maladie Facultative. Exemple du Danemark," par Dr. Johan Kuhn.

[2] In exact figures $995. There were 680 doctors in Jutland, 407 of whom were in general health insurance practice (estimated from *Medicinalberetning for den Danske Stat*, 1934, p. 157 and *Statistisk Aarbog*, 1934, p. 29).

tradition of payment per unit of medical service rendered was a strong one in non-metropolitan Denmark, and at least until recently it has not been surrendered at all willingly. As expressed by the doctors who for years have helped to negotiate the collective contracts for the medical profession in Jutland and in Copenhagen,[3] "Capitation is best for the doctor as for the society and its members. It gives the doctor certainty of income and improves his relation with his patients since the patient is not 'out of pocket' [through payment of 'control tickets'] when he drops in at the doctor's request. It promotes good work on the physician's part and induces care of the little things that lead to serious illness. But the practitioners for the most part do not come willingly to capitation in the first instance. Usually it is the case that the societies push it upon them and after they are obliged to have it, they learn its advantages."

As in England, the points were stressed by medical men that capitation brought the insurance practitioner freedom from detailed record keeping for purposes of bill collecting, gave him an assured income, and, by eliminating the possibility of argument over items of medical service rendered, permitted him to practise his profession in peace and comfort.[4]

It is the opinion of both lay administrators and the doctors active in Danish Medical Association activities, that the trend away from Tariff II (fee system) to Tariff I (capitation) will continue, and that for the medical profession's interests, as well as for those of the insured and the societies, *should* continue, until capitation is the basic system of remuneration in all parts of the kingdom.

As has been previously remarked, the organized medical profession believes the restricted medical list in Copenhagen to be definitely desirable, while they insist upon free par-

[3] Dr. Hjalmer Fridericia in Aarhus, Dr. Johan Kuhn in Copenhagen.

[4] For good statement of these often expressed views, pp. 10-12 of the address referred to in note 3 *supra*.

ticipation at will of all doctors outside the Copenhagen area. They explain that their apparently inconsistent views can be reconciled. The attractions of a metropolitan life are such, they allege, that an undue number of medical men tend to congregate in the capital. They believe that without the check of an insurance list restricted to the number of doctors deemed adequate to meet the needs of the insured patients, serious overcrowding and competition, which would reduce average medical income to an undesirable level, inevitably would result. Free participation in insurance practice in the less populated areas, on the other hand, makes for a distribution of practitioners which is advantageous not only to the insured population but also to the average medical purse.

While it is impossible to find a Danish doctor who objects to health insurance medical service, some practitioners, as might be expected, feel that the capitation amounts and the fees in the "fee schedules" might well be higher. There is, however, comparatively little of such sentiment. The general attitude of the profession is that on the whole they are quite reasonably paid, especially in relation to the circumstances of the insured population and in view of the substantial security and comfort of insurance practice.

Proposed Extensions of Health Insurance Medical Service

Expansion of specialist service in the non-metropolitan areas is the improvement in health insurance "medical benefit" which is most discussed in Denmark.

The National Health Service (Denmark's "Ministry of Health") has in view a program which is based upon a preliminary development of central hospitals having standardized surgical and internal medicine specialists attached as staff. This part of the program was commenced in 1932 and is being pushed along rapidly. By the end of 1936 not only were Copenhagen, Aarhus, and a few of the other larger towns well supplied with hospitals staffed with physicians representing all the special fields of medicine and surgery, but also in more than half of Denmark's twenty-four medical districts there was at least one fairly well equipped central hospital. Attached to these institutions as staff are physicians representative of the major special fields of surgery and internal medicine. It is planned that each district eventually shall have one such central hospital.[5]

It is also hoped in the future to improve rural facilities further by "planting" specialists in certain fields of broad demand, in unequipped areas, as part-time staff attached to the communal hospitals. Such staff would be given the privilege of using their hospital quarters for non-institutional private practice.

It is assumed, also, that staff specialists once thus established would contract with the Health Insurance Societies to give clinic and home treatment to the insured population, thereby adding insurance income to their salary from the hospital and to their fees from their practice among the well-to-do in the locality.

It is believed that by such arrangement, specialists who tend to congregate in the cities could be distributed over the rural areas to their own satisfaction as well as to the advantage of the non-urban families.

Whether or not this will be the exact procedure followed, Denmark's health service tradition and the long emphasized medical benefit of her health insurance system point to a steady improvement of the medical services which will be afforded her rural insured population.

[5] Information furnished by Dr. Frandsen, Director of the National Health Service of Denmark.

PART III: FRANCE

XIX

SUMMARY OF FRENCH HEALTH INSURANCE SCHEME

THE French health insurance scheme went into operation in 1930 as a result of an omnibus measure enacted in 1928.[1] This set up a compulsory system applicable to industry, commerce, and agriculture—not only of health insurance, but also of old-age and invalidity insurance and a limited form of survivors' insurance as well. The French coverage provisions followed the conventional pattern set by earlier compulsory laws in that in general only employees were covered. Homeworkers were classed as employees for purposes of the act, as were the sharecroppers (*métayers*) who comprise a large proportion of France's farming population. Selected groups, such as miners, seamen, railway workers, and civil service employees, for whom special legislation already had provided sickness protection, were omitted.

Voluntary insurance was open to small shopkeepers, farmers, artisans, self-employed non-manual workers, etc., and other persons described as those dependent for a livelihood principally upon their labor, provided their annual earnings did not exceed the limits set for the compulsory insurance.

The main outlines of the compulsory scheme were as follows. The premium was divided between the worker and his employer and, as in most health insurance schemes, the employer was made the collecting agency responsible to the government for both shares. A government subsidy to vary in

[1] Amended in 1929 and 1930. This law does not cover industrial injuries which had been provided for earlier in a Workmen's Compensation Act.

amount with the reduction in poor relief expenditures, as compared with the five-year period just preceding institution of the social insurance scheme, was guaranteed.

Moreover, the French Act, like the Danish (rather than the British measure) which covers all manual employees without reference to their earnings, provided that French workers otherwise subject to the act were excluded without exception from the compulsory provisions if their annual earnings exceeded certain limits.

Both so-called "medical" and cash benefits were provided. The medical benefit, however, differed radically from the usual medical benefit, in that it constituted not actual medical service and treatment but merely partial reimbursement for or advance toward the worker's expenditures on medical, surgical, and hospital care and drugs. This was granted in case of illness of the dependent family as well as of the insured for a maximum period of six months from the date of the first medical consultation in respect of each illness. Through invalidity insurance, the same medical benefit rights were granted for the first five years during which an invalidity pension was drawn. Absolutely free choice of doctor was granted, and it was stipulated that, on principle, the insured should always participate in medical benefit costs. His compulsory share was called the *ticket modérateur*[2] and was intended as a primary check on abuse of medical benefit by the insured.

Revised French Health Insurance Law

The scheme was revised by a *décret-loi*[3] in the fall of 1935. Agricultural workers are now provided for in a separate enactment. Voluntary insurance still is open to small farmers, sharecroppers, etc., in supplement to the compulsory insurance covering agricultural employees.

[2] His actual share was far in excess of the *ticket*, see pp. 174 and 199*ff*.

[3] A *décret-loi* is issued by the Premier, on the authorization of, and to be later ratified by parliament.

As in the original act, the measure relating to non-agricultural labor includes all wage earners in industry and commerce, whether manual workers or non-manual employees who are under sixty years of age and whose annual earnings are at least 1,500 francs ($99) but not in excess of specified maximum amounts.[4] The specified maximum, as in Denmark, is supposed to correspond roughly to the full-time wage of a skilled worker and to vary with the locality and the family obligations of the insured. It ranges from 15,000 to 25,000 francs.[5] Establishment of a money limit in the law itself, however, as is done both in Great Britain and also in France, inevitably creates difficulties with changes in money and price levels. Far preferable is the Danish method of periodic announcement of the insurance income limit by an administrative agency which is directed to vary the figures with the movement of the cost-of-living number. Temporarily, because of nationwide high prices the French level has been placed by a special law at 21,000 francs ($1,386) for the insured without dependent children, and 25,000 francs ($1,500) for the insured with at least one dependent child, irrespective of the area in which the insured person works.[6]

In 1935 there were a little over 10,150,000 workers who, with their families, were entitled to health insurance benefits. These ten million insured were distributed among about 750 health insurance funds.

[4] However, the person drawing an old-age pension, even if under sixty years of age, is not subject to the insurance act, nor are school children who work part time or without salary for their own parents.

[5] For French wage levels at date of original enactment, see pp. 52 and 53 of the 1929 *Bulletin du Ministère du Travail.* In cities other than Paris, workers in ornamental stone carving in 1929 drew the highest wage of 39 francs 31 centimes per day (or just under 12,000 francs per annum, counting 300 days); in the Parisian area, typographical and printing trades drew the highest pay of 54 francs 80 centimes per day (or about 17,100 francs per annum on a 300-day schedule).

[6] The law of August 29, 1936, *Journal Officiel*, 1936, pp. 9290, 9314 (*Bulletin du Ministère du Travail*, 1936, pp. 179-80) indicates that this limit is to be considered a temporary concession to the current low value of the franc and corresponding high level of prices, wages, etc. Francs are converted into dollars at the rate prevailing when the French survey was made—at 6.6 cents per franc.

Aliens, originally covered only after three months' work in France, now are insured on the same terms as workers of French nationality.

The privilege of voluntary social insurance originally afforded to small shopkeepers, self-employed artisans, and professional workers of small means has been abolished. They must now obtain any desired non-commercial insurance through the mutualist fraternal organizations. Voluntary social insurance is now open only to the non-working wife of the insured. She receives *medical* benefits through the insurance of her husband and is allowed to insure herself in the regular insurance fund in which her husband is insured, for *cash* maternity, invalidity, and old-age benefits. For insurance purposes she is rated on the basis of 1,500 francs a year, the minimum insurable salary. Benefits are payable only if she is wholly incapacitated for housework.[7]

The premiums or "contributions" are still divided equally between the employer and the worker. But whereas in the beginning the insured were classified into basic wage groups for contribution purposes, contributions (4 per cent from the worker, 4 per cent from his employer, covering sickness, maternity, old-age, and invalidity) now are based upon the actual wages received.

Benefits of health insurance remain of the same two types, cash and medical. No basic change has been made, moreover, in the medical benefit system of providing merely a certain advance toward, or reimbursement of, the cost of medical and hospital care and drugs.[8] The insured whose dis-

[7] See Article 16, of the 1935 Act, *Journal Officiel*, 1935, pp. 11508*ff*.

[8] The regions of France comprising Alsace-Lorraine retain the health insurance system which was effective during their many years of German affiliation. In this system medical benefit follows the conventional pattern rather than that of the French Act. Medical services are guaranteed and are performed by such doctors as are willing to contract with the funds. Doctors are paid by the health insurance funds on a basis similar to that provided in Manchester until 1927, i.e. an annual capitation (51 francs per insured) is given the doctors collectively; they divide it among themselves according to the work performed.

ability results from deliberate misconduct still receives only medical benefit.

The method used in determining both the insured's eligibility for health insurance benefits and the amount of cash benefit due him has been completely changed, although with approximately the same results as in the original act. Instead of requiring the worker to have made a certain number of contributions (sixty in the preceding quarter), the law now requires his aggregate contributions to constitute a certain amount. Exclusive of his employer's contribution, there must be credited to his account either 30 francs ($1.98) from the previous two quarters or, failing that, 60 francs ($3.96) in the previous four quarters, in order that he may qualify for medical or cash benefit.[9]

The minimum and maximum daily cash benefits, except for a professedly temporary increase of the latter to 22 francs ($1.45), remain at the same levels. The supplemental daily allowance granted for each child under sixteen is still one franc, a little more than the current price of a quart of milk. The only change that has been made has been to permit this allowance immediately on the birth of the child instead of at the end of the maternity benefit period of six weeks.

The cash benefit ranges from 3 to 18 francs a week. In general it approximates the original "half pay," but now is computed according to a schedule based upon the aggregate amount of contributions standing to the worker's credit in the period of reference used.

The earlier proviso that sickness days count for qualifying purposes as contribution days has been replaced by a new section which obligates the fund to place "complementary" contributions to the credit of the insured person whose illness lasts over a month. There is a similar requirement of com-

[9] This has caused such confusion and argument, that many funds in furnishing their members with informative circulars on the new law avoid the necessity of explaining this point by doubling the required amount, i.e. by using 60 francs in lieu of 30 francs, etc.

plementary contributions as part of maternity benefit to the insured woman. Thus health insurance rights are maintained during disability, just as they were under the old system when compensated sickness days counted as "contribution" days for qualifying purposes.

These lump-sum complementary contributions are not only used for computing sick benefit rights, but also serve to keep up the worker's old-age insurance rights, taking the place of the half-rate contributions which were formerly required.

No change has been made in the provisions governing the cash benefit on hospitalization of the insured. He still receives only one-fourth of the usual benefit if he has no dependents, one-half if he has a wife but no children (or aged dependent), and two-thirds if he has both a wife and child (or aged dependent). The waiting period of the original act has been retained.

The protection to insurance rights during unemployment periods has undergone a change in computation technique to make it compatible with the new qualifying and benefit provisions, but the unemployment protection provided is practically the same as before.

An improvement has been made in the situation of the worker who suffers an interruption of work through industrial injury. His employer (or, if insured, the latter's insurance company), in all industrial injury cases causing incapacity of more than a month, must from then on make a flat complementary contribution to the worker's social insurance account of twelve francs (79 cents) a month, i.e. 36 francs ($2.38) a quarter. This keeps up his old-age and invalidity insurance and safeguards his health insurance rights in case of illness immediately following his industrial injury period. Previously, if he fell ill after a non-contribution period because of industrial injury, he was unable to qualify for benefit unless he had voluntarily contributed from his own pocket.

Another liberalizing modification has been made in respect to industrial injury cases. Where formerly the law flatly stated that no health insurance benefits were payable in case of industrial injury, now the health insurance fund must grant benefits to the injured worker whose employer refuses to acknowledge the industrial character of the injury. The worker is required, however, to file suit to enforce his right under the Workmen's Compensation Act, and the health insurance fund is subrogated to the worker's rights in this suit to the extent of the benefits it pays out on account of the industrial injury.[10]

For administration of health insurance, France is now divided into regions instead of departments. All the funds in a region are joined in a union of funds, which serves among other things as a primary reinsurance fund for its members. Behind these unions stands a Central Guarantee Fund.

Mutual Aid Societies, when approved as official insurance funds (*caisses*), still administer both the cash and medical benefits, and supplementary departmental insurance funds are also provided for, to conduct the insurance of persons entitled to insurance who have failed to join an approved society. Each health insurance fund affiliates itself with an invalidity insurance fund. The worker chooses his health insurance fund only.

In summary, it may be said that most of the changes made since initiation of France's Health Insurance Act have been directed at giving the funds clear authority to require compliance with their regulations on pain of loss of benefit, and at the simplification of accounts, bookkeeping, and administration. All the substituted contribution and benefit

[10] In case of non-industrial accidents caused by negligence of a third party and giving rise to a right to sue for damages, insurance benefits are payable, but the fund is subrogated to the worker's rights, to the extent of benefits paid. No compromise of the claim against the third party is good against the fund unless the fund has had fifteen days' notice by registered mail. The worker (but not the fund to the extent of its interest) loses his right to sue unless he notifies the third party that he is insured and names his fund.

provisions of the 1935 revisory act were intended to approximate those put into operation in 1930 but with a simpler computation technique. There seems to be no movement of strength to exchange the medical benefit arrangements adopted in the original social insurance act for the conventional social insurance medical benefit which guarantees the actual medical services themselves to the insured families.

XX

HEALTH INSURANCE MEDICAL BENEFIT

"MEDICAL benefit," reads Article VI of the French Health Insurance Law, "covers the expenses of general and specialist medical care, the expenses of medicines, drugs, and medical and surgical appliances (including essential artificial dentures), hospitalization costs, the cost of treatment in sanatoria, and transportation as well as the cost of surgical operations necessary for the insured, his spouse,[1] and the children dependent upon either of them."

This rather ambitious provision is radically limited by subsequent paragraphs of the section. The upshot of these modifications is that only a varying fraction of the listed costs of illness is covered by the insurance.

The terms and conditions which control the size of this fraction and the procedure for obtaining it are traceable directly to the organized medical profession of France. The original French Health Insurance Bill provided the conventional health insurance medical benefit, i.e. it guaranteed the medical services themselves. This was defeated through the efforts of the French Medical Association, which introduced the substitute bill which was enacted into law.[2] The terms and conditions of French medical benefit were designed to maintain the following principles, which, the Medical Association maintains, comprise the French "Medical Charter":

[1] A spouse who is also a wage earner, if earning more than the limits set for compulsory health insurance, has no right to medical benefit.

[2] See Antonelli, *Le Droit des Assurances Sociales*, Tome I, pp. 6-7, 30th lesson, for a full account. Also, cf. Article 4, 1928 Act, with Article 4, Amendatory Act of 1930.

1. Absolutely free choice of doctor.
2. Payment of the doctor in an amount determined between the individual doctor and patient.
3. Payment according to unit of service rendered.
4. Payment of the doctor directly by the patient.
5. Preservation by the doctor of complete secrecy with regard to his professional knowledge of his patient's condition as "privileged" information which the doctor will never communicate.

Thus the doctor who treats an insured patient in France does not receive remuneration either from the insurance organizations, as does the insurance doctor in Denmark,[3] or from the governmental authorities as does the health insurance doctor in Britain. He serves his insured patient as he always has done on the terms which he deems acceptable. His patient, through his medical benefit, then receives from his insurance fund a fraction of the doctor's bill; this benefit is given to him when he furnishes his fund with an appropriate slip signed by the doctor. The portion of the physician's bill which is met or "reimbursed" by the insurance fund is not even a definite proportion of such bill.[4] It is merely 80 per cent[5] of the amount set in the fund's *tarif de responsabilité* as the appropriate fee for the type of medical service rendered. The *tarif* is a schedule (20 per cent, termed the *ticket modérateur* or controlling ticket, is not reimbursed) which lists the various types of medical and surgical services and values each for reimbursement purposes at a certain number of francs. The doctor uses mere symbols in the insurance slip which he signs for the insured. These symbols indicate that the service he performs falls into a certain price category of

[3] See pp. 207-8 and 242*ff*. for illustrations of some voluntary arrangements in contravention of this general situation.

[4] The original provisions of the "doctor's bill" stipulated that the fund should pay 80 per cent of the medical expenses evaluated at the usual prices charged in the neighborhood. This was amended before the act went into effect, to provide the present scheme.

[5] Originally it was 85 per cent for the lowest paid workers.

the *tarif*[6] but do not reveal the nature of the actual medical service itself nor the nature of the ailment of the patient.

On pain of forfeiting the right to assistance from the Central Guarantee Fund in case of deficit, the fund's *tarif* amounts must not exceed the maximum limits of the *tarif de réassurance* set by the Ministry of Labor. This latter reinsurance schedule is set by the Minister of Labor after consultation with his Advisory Administrative Social Insurance Council (*Le Conseil Supérieur des Assurances Sociales*[7]) and taking account of the Minimum Fee Schedule of the French Medical Association.[8] The Social Insurance Council is a large unwieldy body with a membership representing each of the interested ministries, both houses of parliament, the representatives of each type of insurance fund, and each of the professional and technical groups: pharmacy, medicine, midwives, dentists. Its actual technical work is performed by a permanent section chosen from its membership by the Council at large and functioning chiefly through three subsections into which it is divided: (1) Financial, (2) Administrative and Judicial, (3) Medico-Pharmaceutical.

The *tarifs* of the individual funds are established in the first instance by the funds and incorporated, together with regulations about presentation of claims for medical benefit, in collective agreements between the funds and the medical associations in the different regions. The *tarifs* seldom greatly exceed the reinsurance schedule's rates and, since they must be approved by the social insurance authorities to be effective, they are not much lower than this schedule, ex-

[6] See pp. 199*ff.* for further discussion of this *tarif.*

[7] This body is hard to classify. To use the words of an authoritative commentator, Étienne Antonelli, professor of law and specialist in labor law, "It is a little more than an advisory body and a little less than an autonomous body 'acting of its own initiative,' " *Le Droit des Assurances Sociales*, Tome II, p. 212.

[8] Similarly, the minimum fees of the other technicians, pharmacists, dentists, nurses, etc., are to be consulted in setting *tarif* rates for their services. In this discussion, when the term Medical Association is placed in capital letters, it refers to the national organization. When otherwise, a local association is meant.

cept when the fund's financial situation is so weak as to necessitate approval of such rates. Medical interests, of course, press toward high rates. There must be approval of each *tarif* by the regional social insurance commission or, failing such approval, by the permanent section of the Social Insurance Council to which the funds may appeal against regional refusal. The representative of the Ministry of Labor who sits with the regional authorities, may appeal against an approval by the latter which he deems unjustified. The regional social insurance commission includes representatives in equal proportions, (1) of the Ministry of Labor and the Ministry of Health, (2) of the funds, and (3) of the doctors and other technicians.

The reinsurance schedule since 1934 has divided the regions of France into three divisions, according to their population and degree of industrialization. Division A includes cities with more than 500,000 population and such highly industrialized areas as the Seine and the Rhone. In Division B are the cities which are not in Division A and yet have a population of more than 100,000. All other regions are classified as Division C. The basic office-consultation rate in this schedule is supposed to approximate one-half the average daily wage of skilled labor. This is roughly the case, except for the Paris region where it is equivalent only to half the average daily wage paid semiskilled workers.

The 1936 reinsurance *tarif* fees were as follows:[9]

Office Consultations (cited as C)

Division A	18 francs ($1.19)
Division B	15 francs (99 cents)
Division C	12 francs (79 cents)

House Calls (cited as V) office-consultation rate plus mileage.

Mileage—(1) a flat rate of 2 francs in cities (including their suburbs) of 100,000 or more; (2) in regions other than (1), 75 centimes per kilometer (a little less than a mile) in non-

[9] Rates were advanced at end of 1938. See note 23 *infra*.

mountainous regions; 1.25 francs in more mountainous regions (minimum 2 francs).

SUNDAY CONSULTATIONS AND HOUSE CALLS—add 30 per cent to the rates above.

NIGHT CONSULTATIONS AND HOUSE CALLS—add 100 per cent to the rates above.

When more than a mere consultation is involved, the general practitioner's services are designated in terms of multiples of the basic consultation C and rated accordingly. Thus, for example, in the present reinsurance *tarif* in Division A (metropolitan area) C = 18 francs, and the following rates apply:

Nature of Service	*Reimbursement Value*				
Inoculation with a preventive serum	C 1.2	=	21.6	francs	($1.43)
Dressing a serious injury	C 2	=	36	francs	($2.38)
Hernia reduction	C 3	=	54	francs	($3.56)
Lumbar puncture	C 4	=	72	francs	($4.75)
Giving a general anesthetic	C 6	=	108	francs	($7.13)

Official lists of general practitioner services and of those deemed surgeons' and specialists' services have been compiled in collaboration with the Medical Association and promulgated by the Ministry of Labor. The latter two types of services are designated in the *tarifs* in terms of K and a digit by which it is multiplied. The symbol K signifies in all regions in respect of services to a non-hospitalized patient, reimbursement at the rate of 8 francs when the digit is less than 40, 15 francs when it is 40 or more. For instance:

Nature of Service	*Reimbursement Value*				
Removal of tonsils and adenoids	K 20	=	160	francs	($10.56)
Hernia operation	K 40	=	600	francs	($39.60)
Removal of appendix	K 60	=	900	francs	($59.40)

As previously stated, agreements between the funds and the medical associations must be approved by the regional social security authorities or, on appeal, by the Social Insurance Council. If no collective contract is agreed to, the fund, in lieu of reimbursement in accordance with the usual schedule, may pay the insured a small flat-rate lump sum[10] (fixed in a special *tarif*) for all medical services, regardless of their nature. The latter situation is most exceptional, as it is disadvantageous to the doctor as well as to the insured, in that the smaller reimbursement rights of the insured press him to cut down on medical care. In view of this effect on medical practice, doctors see to it that collective contracts are made, except when this is impossible, i.e. when there is no "certified" medical association in the region. Certified means properly representative of the doctors practising in the locality. To achieve this status, a local medical group must first seek recognition by the National Medical Association as a representative local group and then be approved by the Social Insurance Council. If a local group fails to obtain the approval of the national professional group, it may appeal to the medical-pharmaceutical section of the Social Insurance Council, which, it will be remembered, is one of its three main subsections. This subsection includes at least two physicians and one pharmacist. If it agrees with the National Medical Association that the petitioning local medical group is not representative, the matter is ended. If, however, the subsection believes the local group has a legitimate claim, certification may be granted by the Social Insurance Council despite the National Medical Association's opposition.

The right to reimbursement toward medical bills extends to bills paid to any medical man authorized to practise in France. There is absolutely free choice of doctor. Not even the medical association itself in collective agreement with the funds may intrude on this principle.[11]

[10] See Article 37, sec. 9, of the 1935 Act.

[11] The original executive decree of 1929 setting up the administrative regulations for social insurance purported to limit this choice of physicians within the locality. Article 32, Décret de 5 Avril, *Journal Officiel*, 1929, p. 3,974. This decree never was put into legal operation and the limiting provision was omitted from the 1930 decree which set up the rules actually used when the act began to function. Article 18 of the Décret de 25 Juillet 1930, *Journal Officiel*, 1930, p. 8,453.

Thus in a case litigated to the French Supreme Court, it was held that a proviso in the collective agreement between the doctors and the funds that only bills from doctors domiciled in France would be honored for medical benefit was violative of the free choice guarantee. The patient was allowed reimbursement on a bill for medical services rendered in Geneva, the court suggesting that had the fund alleged the impossibility under the circumstances of control of the patient to check fraud, the decision might have been otherwise.[12]

In deference also to "free choice," the decree setting up the official administrative regulations[13] provides that the insured may get reimbursement (although only on the flat-rate *forfait* basis which is used in the absence of collective agreement) by presenting receipted medical bills, even when his doctor is unwilling to abide by the collective agreement between the funds and the medical association. Such refusal apparently occurs very seldom and is related to the doctor's desire to avoid subjecting himself to the medical association's disciplinary machinery. All collective contracts provide that the doctors agree to submit to such disciplinary procedure, and governmental Official Regulations[14] state that a doctor, even when not a member of the medical association, is deemed to have agreed to the terms of the collective agreement when he signs insurance slips to enable his insured patients to get reimbursement benefits.

The purely private practice relationship of doctor and patient as to the amount of medical charges, i.e. that the size of the bill is set in each instance by the doctor, has also been construed by the central administrative body to be guaranteed by the *libre choix* (free choice) of the French act, even as

[12] *Caisse Primaire Départementale des Assurances Sociales de la Haute-Savoie contre Dame Perrissin, Dalloz, Rec. Heb.*, 1934, p. 284.

[13] Décret du 19 Mars 1936 (*Journal Officiel*, 1936, p. 3,233, or see *Bulletin du Ministère du Travail*, 1936, p. *33). *General Administration of Social Insurance*, Article 10, sec. 2. For discussion of disciplinary control see pp. 215*ff*.

[14] *ibid.*, Article 10, sec. 1.

against collective medical action. Thus the provisions in two collective agreements between the funds in two departments and the corresponding medical associations, to the effect that all doctors practising among insured families would charge "not less than" (in one agreement) and "not more than" (in the other agreement) the minimum fees of the medical association's schedule, were denied approval by the Social Insurance Council as violating the spirit of "free choice."[15]

The drug benefit which is included in French medical benefit gives the insured family partial reimbursement for drugs, appliances and therapeutic supplies. For medicines proper, a fraction of the actual cost and not of a *tarif* amount is granted. Their cost, however, is regulated by collective agreements between the funds and the pharmacist's associations. Within medical benefit, however, i.e. if reimbursement is expected, the attending physician is no longer permitted absolutely free exercise of his discretion in prescribing for his insured patient, as he was in the original act. There is now reimbursement of only one prescription for each medical consultation. The physician, moreover, is specifically enjoined in the revised act that he has "the liberty of prescribing only according to the most strict economy consistent with proper treatment" of his patient's condition. Reimbursement allowed on the cost of prescriptions is 80 per cent of the prescriptions which do not involve over twenty-five francs ($1.65) and 60 per cent of those of higher cost.[16] Even with the higher priced prescriptions, however, 80 per cent reimbursement is permitted on specified serums[17] and, with the approval of the Medical Superintendent of the supervisory *Contrôle Médical* of the fund, in cases of any disease necessitating very expensive drugs.

[15] Agreements made in Finistère and in Belfort. Antonelli, *op. cit.*, Part IV, Chap. II, sec. 2.

[16] Originally a straight 85 per cent of the cost was reimbursed.

[17] Serums having an organic base and regulated by the law of June 14, 1934.

Both the original and revised acts authorize the Minister of Labor, on recommendation of his Social Insurance Council, to set special reinsurance schedule limits for reimbursement of proprietary or patent medicines. So far this privilege has not been exercised, although commissions have been formed at two different times (the last in 1937) to draw up such a schedule. Great caution in the use of proprietary medicaments is constantly urged by the medical journals of the organized profession—in the hope that voluntary restraint will forestall limitations on free prescribing by the doctor.[18]

For medical and surgical appliances and supplies, the fund sets a schedule,[19] with 80 per cent reimbursement as in the case of medical fees, and, on authorization of the attending doctor, the cost of auxiliary services such as may be rendered by a nurse or a masseur also will be reimbursed at specified schedule rates.

The French act always has been ambiguous in respect to the choice of hospital permitted the insured. It is clear that any public hospital, and also such private hospitals as have contracts with the fund, may be entered. It is not clear whether care in other institutions is within medical benefit. Reference to official regulations, however, makes certain at least what the administrative authorities contemplate.

Until the middle of 1936, reimbursement toward the cost of institutional care was permitted by administrative regulations only when the insured patient entered either a public hospital or a private hospital with which his fund had contracted. If the insured wanted to be institutionalized elsewhere, he received reimbursement only for the medical or

[18] The Minister is now also empowered to put a financial limit on the drug benefit payable per each *feuille de maladie* (insurance record slip).

[19] Except for such ordinary supplies as cotton, gauze, adhesive tape, etc., unless a physician is licensed as a dispensing doctor, in furnishing medicaments or surgical supplies and appliances, he is guilty of unlawful practice of pharmacy. In practice, however, all physicians keep on hand medicated bandages, iodine, etc., to use in their practice. The doctor then gives his insured patient a prescription order for the amounts used. The latter fills the order and gives the materials to the doctor to replenish his supply.

surgical fees involved. Present regulations, however, provide that the insured may enter any institution "which accepts the *contrôle* rules of his fund," i.e. which agrees to comply with the administrative rules which are directed at supervisory control of the insured person who claims benefits.

This would include, of course, even commercial institutions; however, as the cost of a bed in a commercial private hospital or "nursing home" (*maison de santé*), plus medical fees charged, is far beyond that of a hospital proper, this change has no great practical importance. The insured is not likely to wish to enter such an institution. In recognition of this fact, the organized medical profession, so far without success, have been driving for a more liberal reimbursement for "nursing homes" as essential to the "free choice" principle.[20]

For institutionalized cases the reimbursement arrangements include two elements, one allocated to hospitalization itself, the other to medical and allied services of the staff other than the salaried nurses. The bed-day *tarif* rate signifies the actual amount payable by the fund to the institution toward the cost of the bed to the insured person. The patient pays no *ticket modérateur* (i.e. the usual 20 per cent share). The reinsurance schedule sets an individualized rate for each hospital, which is a percentage (in 1936, 70 per cent)[21] of the amount payable by the Medical Assistance authorities for their cases. The schedules of the individual funds vary these rates by special contracts with the various hospitals. As in the case of medical care, what the insured person shall actually be charged for hospitalization is, of course, not regulated by the health insurance measure. The contracts made by the funds with the various hospitals, however, stipulate what the insured shall be charged by the institution as well as the reimbursement rate to which the fund agrees.

[20] For reference to this situation in the official medical journal, see *Le Médecin de France*, April 1937, pp. 334-6.

[21] Raised to 80 per cent in 1937.

In Paris all hospitalization of insured families is handled by the Public Assistance authorities acting under one central agreement with the health insurance funds of the area. The per diem cost of a bed in many of the hospitals is more than double that of the *tarif* amount contributed by the fund, leaving a heavy burden upon the insured. This in turn means a burden upon the Public Assistance authorities, which must provide for the insured who are unable to pay their complementary share.

Rural hospital bed-rates on the whole are far below those of metropolitan institutions, and institutional care is in consequence more completely an insurance benefit in the non-urbanized areas than it is in the cities.

This is illustrated by the table below, which reproduces the contribution (*tarif*) rates offered by a well run rural de-

	Medical Bed			Surgical Bed		
Name of Hospital	Daily rate charged the insured	*Tarif* rate paid by the funds	*Tarif* rate which would have been payable by fund in absence of contract (reinsurance rate)	Daily rate charged the insured	*Tarif* rate paid by the funds	*Tarif* rate which would have been payable by fund in absence of contract (reinsurance rate)
	francs	francs	francs	francs	francs	francs
Coullomiers	23.69	20	15.79	25	24	16.81
Sagny	20	20	13.87	22.41	22	14.93
Melun	23	20	15.63	26	24	17.36
Montereau	21.76	20	14.50	24.23	24	16.15
Moret	18.65	18	12.43	18.66	18	12.43
Brie-Compte-Robert	18.10	13	12.06	18.94	16	12.62
La Ferte Gaucher	13.29	13	8.86	14.94	14	9.96

partmental fund (*Caisse Départementale de Seine et Marne*) for hospitals with which it has made contracts. These rates are substantially in excess of those of the reinsurance schedule and in most cases approximate the actual cost of hospitalization to the insured.

As stated above, in addition to a bed-day rate, the hospitalization reinsurance schedule sets rates payable for medical services rendered by the hospital staff. These medical honoraria are reimbursed like ordinary medical bills as to 80 per cent only of the *tarif* rate, a 20 per cent *ticket modérateur* being left to the insured. The reinsurance schedule rate for these honoraria is a straight nominal *forfait* of 4 francs a day for both medical and surgical cases in all public hospitals in medical teaching centers. In contracting with other institutions, while the same rate is set for medical cases, the fund is permitted a choice of two rate procedures for surgical cases. Either the usual *forfait* of 4 francs a day or a specified lump sum for the operation (including twenty days aftercare) may be elected. The lump sum is to be computed by referring to the symbol rating of the operation in the *tarif* or the medical association's fee schedule,[22] using 3 francs as the equivalent of K if the coefficient is not more than twenty-four, and 6 francs as the equivalent of K when the coefficient is twenty-five or more.[23]

To illustrate, an appendicitis operation is K 60 in the surgical *tarif*. The lump-sum honorarium for such an operation in a hospitalized case involving a twenty-day stay is, therefore, $6 \times 60 = 360$ francs. If complications set in, and the patient spends another two weeks in the institution, 4 francs a day is added to the honorarium, totalling for the entire case 416 francs.

The daily *forfait* basis produces much lower honoraria. A twenty-day appendicitis case calls for only 80 francs, and an

[22] The same symbols are used in both.

[23] In December 1938, all rates advanced. These became 3 fr. 50 and 7 fr. 50 respectively. See *Journal Officiel*, 1938, pp. 11791*ff*.

operation entailing thirty-four days' hospitalization calls for an honorarium of only 136 francs, or about 33 per cent of the lump-sum *tarif* rate.

All medical honoraria are paid directly to the hospital, which distributes them among the staff physicians in accordance with the hospital's rules.[24] Thus in case of hospital doctors, the French Medical Charter's principles of payment according to "direct contract" between doctor and patient and payment "directly to the doctor" have been abandoned.

Honoraria (in addition to hospital charges) for medical services rendered by the staffs of public hospitals and philanthropic private institutions are recent developments in France, as in the "voluntary" hospitals in Britain. Traditionally, hospital care in France included medical and surgical services, and staffs of these hospitals, as in the "voluntary" institutions in Britain, were either wholly unpaid or in receipt of nominal honoraria from the institution. The smallness of the *forfaits* demanded in respect of insured patients for the services of such hospital staffs is understandable in the light of the history of these institutions.

It seems recognized that a large fraction of the socially insured population would be hospitalized through the Public Assistance authorities were it not for social insurance, and it appears to be generally accepted in medical circles that relatively little can be expected to be paid the hospital staffs on the hospitalization of such insured patients. It is contended by the medical associations, however, that there is a substantial

[24] The medical honoraria for health insurance cases in Paris hospitals are divided in accordance with the rules made by representatives of the medical staffs, in the proportions agreed upon annually. In the 1935 report the proportions were as follows:

Regular staff consultants	46 per cent
Assistants in charge of laboratories, etc.	15 " "
Internes	13 " "
Student nurses	11 " "
Midwives	10 " "
Medical student assistants (other than internes)	5 " "

See "L'Application des Assurances Sociales dans les Hôpitaux," *Administration Générale de l'Assistance Publique à Paris* (1933), p. 3.

better-incomed fraction of the insured who are proper potential clients of the private commercial hospitals. The doctors maintain that such persons should be encouraged to choose their surgeons freely and that, in justice to the members of this group, there should be a much more generous reimbursement *tarif* for medical and surgical services in a private *maison de santé* (*nursing home*) than for those rendered in public hospitals.[25]

Procedure Required in Obtaining Benefits

Since the 1935 revision of the Social Insurance Act, right to all medical benefit has been absolutely dependent upon conformance with requirements designed to promote effective supervision by the fund and to check fraudulent imposition, i.e. to strengthen *contrôle médical.* On pain of loss of all benefit, both cash and medical, the first medical consultation must be reported to the fund within three days. There are exceptions from this harsh rule, for unavoidable non-compliance.

A sickness folder, *feuille de maladie*, must be obtained by the insured in order that he may receive either his cash or medical benefits. This *feuille* or insurance record slip is a printed form which (1) contains space for the data about the insured's conditions which are required for establishing his rights, and (2) also gives full instructions about his rights and the steps which he must take to obtain them.[26] Either a de-

[25] The suggestion was made to a special study commission (representatives appointed by the medical federation and by the funds) by its medical members, that the symbol K should be valued at 18 francs for operations performed in *maisons de santé* (instead of 3 and 6 as in the hospital *tarif*). Even this suggestion was objected to by some physicians as "strangling" direct arrangements between doctor and patient. (See *Le Médecin de France*, April 1937, pp. 335-6.) In the minimum fee schedule of the medical association of the Seine, K equals 20 francs.

[26] The latter is required by Article 113 of the *Décret de Règlement d'Administration* pursuant to a section in the revised act, which resulted from a decision of the French Supreme Court holding that a woman could not be denied nursing benefits on the ground that she failed to observe the regulations set up unless she had been personally notified of the rules. "*L'Affaire Époux Hardy,*" *Cour de Cassation Civile*, *Dalloz*, *Rec. Heb.* 1935, *Sommaire*, p. 5.

tachable "notification of illness" card is annexed or a separate one enclosed. This notification card, which must be mailed to the insured's fund, carries the postal franking privilege, a privilege granted all social insurance communications sent to or by a fund. As in all health insurance schemes, cash benefit is payable only on medical certification by the attending physician, and reimbursement of medical bills depends on medical attesting of the bill's character. (Cash benefits are reduced on hospitalization, in accordance with a sliding scale which varies with family responsibilities.[27])

The first *feuille* is good only for a maximum of eight days. If illness continues, a second record slip must be obtained or benefits will be forfeited. This, too, serves only for an eight-day period, each subsequent record slip being effective for fifteen days. As previously stated, the doctor fills in the data necessary for the insured's reimbursement rights, in symbols which preserve complete secrecy as to the nature of the doctor's actual services and the patient's ailment. (The same symbols are used in both the medical association's fee schedule and in the fund's schedule for the same service.) The symbol used by the doctor when reporting on a case may mean any one of many services valued in the schedule at the same rate. For example, in the *tarif* of Paris and its environs, C 1.6 means any one of the following services:[28]

1. Taking a specimen of blood for laboratory examination.
2. Giving a single intravenous hypodermic treatment (involving anything but arsenobenzine).
3. Single antitetanus inoculation plus a simple dressing of a minor wound or cut.
4. An average dressing or bandaging case.
5. Subcutaneous hypodermic treatment.
6. Arsenobenzine inoculation when one of a series.

[27] See p. 170 *supra* for details.

[28] *Year Book of the Medical Association of the Seine*, 1935, p. 78.

7. Curative (not preventive) serum inoculation.
8. Taking more than three stitches—or less than three stitches if linen or horsehair is used.
9. Removal of foreign substance from the body (if an easy and uncomplicated case).
10. Dressing the antrum.
11. Lancing a boil (with or without injecting novocaine).
12. Removal of a surface abscess or tumor.

Ordinarily the fund expects the insured to pay his doctor bill and then apply for the amount of reimbursement allowed by the schedule. However, regulations permit an advance of this amount in lieu of reimbursement when the worker makes the appropriate application proving his need of such an arrangement. The advance is made for only one item at a time. No further advance is made except on proof that the amount has been paid to the doctor. When this is established, the fund will give an advance toward the next instalment on the bill.

When asking for reimbursements (as for other health insurance benefits), the insured is supposed to produce his receipts for contributions paid in the qualifying reference period. The employer, on paying the contributions, gets a receipt in duplicate. One set he is obligated to furnish to his worker at least every three months. If the insured is unable to furnish the regular receipts but can produce an official labor record book kept by his employer, showing wages earned and deductions made for social insurance contributions, his benefits are granted on this evidence.[29]

If the worker falls ill when he is away from home, he can obtain his health insurance benefits through an affiliated fund. Such affiliations are required in every region, and full

[29] If the fund, under such circumstances, finds on checking the records that the contributions indicated by the labor record book have not in fact been made, it can pursue the employer in question for benefits unjustifiably paid.

information on affiliations can be obtained by the insured from the local departmental fund.

The insured patient is legally obliged, as under the British and Danish schemes, to seek his medical care at the doctor's office if his physical condition permits. According to regulations contained in or based upon the administrative *décret*, the insured must obey his physician's orders, particularly with respect to staying in bed, refraining from exercise, etc. He must abide by the regulations of his fund, which to be effective require approval of the Ministry of Labor. (All such regulations are binding upon him only when he has been notified of their content.[30]) For disobedience the funds provide penalties of 10 per cent to 25 per cent reduction of benefit, punishment contemplated by the administrative authorities and provided for in one of the clauses of the "Model Rules" promulgated by the Ministry.

Since the 1936 regulatory *décret*, the insured has been required to get an authorization or "acceptance of responsibility" from his fund when he wishes a consultation of doctors, the services of a specialist, a surgical operation, or auxiliary treatment such as massage or medicinal baths.[31]

If the insured person goes to a specialist for services listed as suitable for performance by a general practitioner, reimbursement is merely at the general practitioner rating.

When a house call entails treatment of several patients who are members of one insured person's family, reimbursement is on the basis of a house call for only *one*, and on an office consultation basis for the other patients treated.

The insured is required, on pain of loss of benefit (both medical and cash) to permit full *contrôle médical*, i.e. investigation of his condition, through personal interview and

[30] This rule was incorporated in the act itself (Article 31, sec. 4, *Journal Officiel*, 1935, p. 11,602) because of the decision of the *Cour de Cassation* (the French Supreme Court) referred to in note 26, that violation of administrative rules could not result in penalty unless the rules were brought to the personal attention of the insured. Ergo funds provide such notification on the insurance record slips (*feuilles de maladie*).

[31] The insured gets written notification of this rule on his first *feuille de maladie*.

medical examination by the supervisory administrative staff of his fund.[32] If his condition permits, he must come to the fund's headquarters for this investigation. If it does not, he must receive the fund's inspector, be it doctor, nurse, or lay visitor, at his home. He may have his own physician present at a *contrôle* examination. All disputes between the insured person and the fund over the state of his health and his consequent right to benefits go for first and final settlement to a technical commission of three, composed of the insured's doctor, a doctor representing the fund's *contrôle médical*, and a third physician appointed by the presiding justice of the trial court.

One purely preventive medical benefit is frequently given, i.e. the right to reimbursement for a complete medical checkover every five years. This, the law specifically states,[33] may be made the privilege of both the worker and his spouse and the children in his charge, "whether sick or well." It is not, however, required.

Dental Benefit

Reimbursement toward the cost of general dental care is given in accordance with established fee schedules which, as with other *tarifs*, usually corresponds in general to the central reinsurance schedule. A special dental *feuille* must be obtained from the fund and filled out by the dentist. As in Denmark, health insurance in France is promoting long-overdue care of the teeth of the general population. Some of the departmental and large mutualist funds now employ their own staff of dentists and have thoroughly modern equipment.

Contribution toward the cost of artificial dentures is provided only under two circumstances: (1) absolute necessity on health grounds, (2) necessity on personal-appearance

[32] For full discussion, see pp. 214*ff*.

[33] The original act made this a required benefit, but for the worker exclusively. The 1935 revisory act made it optional with the fund and included the family as possible beneficiaries.

grounds because of the occupation of the worker. For the resolution of arguments over this benefit, a special commission of three must be set up within five days of application by the insured. The commission comprises one representative of the fund in question, one doctor appointed by the local medical association (or associations[34]) with which the fund has contracted, and a dental surgeon appointed by the association of dental surgeons with which the fund has contracted. If contracts have not been made with either the medical or dental associations, the presiding officer of the local civil court makes the appointment. The commission's decisions are final and must be handed down within twenty days.

Maternity Benefit

As in both Denmark and Britain, pregnancy and childbirth are specially provided for in the French health insurance scheme. Cash benefits for a three-month period (six weeks before and six weeks after confinement) are granted the insured woman, on condition that she cease remunerative work during the compensated period. These "indemnities of repose" are in the amount of the regular cash sick benefits. A woman who is insured in her own right is also entitled to a cash nursing benefit if she nurses her infant. This must be at least 175 francs ($11.55) a month for a period of four months, and the maximum which the fund may provide is 850 francs ($56.10). If the woman is not able to nurse her child (medical certification of this fact is required), she may obtain "milk bonuses," which are 50 to 60 per cent of the nursing benefit amount.

A reimbursement grant for the cost of medical services and drugs needed during confinement is made in the case of both an insured woman and the wife of an insured man. Originally, i.e. from the inception of the act in July 1930 until 1936, reimbursement of medical bills for obstetrical services was on

[34] In some regions the specialists and the surgeons have their own separate organizations.

the same basis as reimbursement of bills incurred for illness, i.e. 80 per cent of a *tarif* amount. As a result, however, of the revision enacted in the fall of 1935, the participation of the insured in the *tarif* amount, i.e. the *ticket modérateur*, has been abolished. The fund now sets an inclusive lump sum (*forfait*), which is paid for the medical and pharmaceutical expenses of the entire case.[35] This includes delivery of the child and normally required subsequent medical attention, whether delivery is made at home or at the hospital. This is payable, however, only upon condition that the applicant (1) reports to her fund the fact of her pregnancy (established by either a doctor or a midwife) at least four months before the anticipated date of confinement, and (2) complies with her fund's regulations relative to prenatal and postnatal care.

These conditions have been written into the law itself because of a series of decisions of the French Supreme Court holding that funds did not have the privilege of refusing benefit because of non-compliance with their rules directed at enforcing prenatal supervision.[36]

The patient is allowed free choice of any midwife or doctor. If she wishes to be hospitalized, the fund pays to the hospital the proportion of the *forfait* corresponding to the share

[35] This was added in Article IX, sec. 7, of the revisionary *décret-loi* of October 28, 1935. The reason for a controlling fee payable by the patient obviously is absent in maternity cases and the abolition of the *ticket* was in the nature of correcting a mistake in the original act.

[36] The funds were directed to pay benefits despite the breach of rules in the following cases: *Schreiber v. Caisse Départementale, Cour de Cassation Civile, Dalloz, Recueil Hebdomadaire*, 1932, p. 473 (did not report pregnancy until two months before confinement); *Caisse Départementale v. Depuis, Cour de Cassation Civile, Dalloz, Rec. Heb.*, 1933, p. 283 (did not report pregnancy until three weeks before confinement); *Caisse Départementale vs. Époux Boyadjian, Cour de Cassation, Dalloz, Rec. Heb.*, 1935, p. 113 (did not report pregnancy until two months before confinement); *L'Affaire Premy, Cour de Cassation Civile*, April 4, 1934, *Dalloz, Rec. Heb.*, 1935, *Sommaires*, p. 5 (did not report that she had ceased employed work six weeks before confinement); *L'Affaire Époux, Hardy, Cour de Cassation Civile*, January 16, 1934, *Dalloz, Rec. Heb.*, 1935, *Sommaires*, p. 5 (did not observe month consultation and certificate requirements demanded by rules for nursing benefits).

which the hospital provides of the total medical care and drugs involved in the case.

If a pathological situation develops (i.e. disease as distinguished from abnormal parturition), ordinary medical benefit becomes applicable and it is granted, if this occurs prior to confinement, for a full six months after childbirth, after which invalidity insurance applies as in any sickness case.

Most of the departmental funds, as well as the larger mutualist funds, give special additional maternity benefits, such as reimbursement for several prenatal medical examinations or arrangement that the latter shall be made by a dispensary at no cost to the insured. For example, the departmental fund in the Seine et Marne region,[37] in addition to the regular maternity *forfait* pays a special benefit of 40 francs ($2.64) for two prenatal medical examinations, with an added 20 francs ($1.32) for laboratory analyses in connection with the first medical examination, whenever the insured cannot obtain such an analysis without cost from a regional dispensary of the Bureau of Social Hygiene. To encourage hospitalization, the fund also allows unreduced cash benefits for a maximum of twelve days in the hospital.

All of the larger funds in France, and many of the unions of funds, now provide well organized prenatal and postnatal supervision, either by a directly created visiting nurse maternity service or by collaboration with existing services maintained by social hygiene dispensaries or by medical schools and nurses' training schools.

The union of funds in Lyons[38] has contracted for such supervision with the Franco-American Foundation, an organization initiated during the World War by the Children's Bureau of the American Red Cross. This Foundation contracts with obstetrical specialists who do prenatal examina-

[37] See pp. 201-2, n. 4, for discussion of *tarif* rates and pp. 239-41*ff*. for discussion of maternity service of Nancy.

[38] Started under the directorship of Dr. William Palmer Lucas, (San Francisco Pediatrician, who was Director of the Red Cross' Child Welfare Section in France).

tions and conduct baby clinics for all the funds in the region. The physicians are paid on a session basis,[39] the session being a three-hour period. The session expense is divided among the insurance funds according to the number of their members who are examined. The insured pays nothing for these examinations.

In addition to its share of the session expense, the insurance fund pays 25 francs for each pregnancy case supervised, in return for which the Foundation guarantees an unlimited visiting nurse (*infirmière*) service during the prenatal and postnatal periods. *Infirmières* visit the cases in their charge as often as necessary, in some instances almost daily, without any extra cost to the funds.[40]

Duration of Benefits

The ordinary right to reimbursement is limited (as is cash benefit) to six months for any one incapacitating illness.[41] A relapse, after an apparent recovery, counts as the same illness if it occurs within two months. If it does not occur within two months, it is counted as a new illness giving rise to a new six months' period of medical benefit, provided, however, that the apparent recovery is reported to the fund within eight days.

Medical benefit, at a reduced rate of reimbursement if the fund wishes, is extended to at least two additional years when the insured continues to work while taking medical treatment. Similarly, if he is in need of special preventive medical treatment designed to forestall chronic illness or invalidity, the insured is entitled to a contribution toward the

[39] 150 francs per three-hour session.

[40] The districts supervised by this foundation are the only ones in the Rhone area in which the mortality rate does not exceed the birth rate. In 1935 the infant mortality rate for the districts which it supervised was 2.29 in contrast to a rate of 5.96 for the entire area. For the Nancy maternity service, see pp. 239*ff.* and see p. 217, n. 3, *in re* certification of *infirmières*.

[41] The worker whose incapacity to work through illnes lasts over six months is entitled to invalidity insurance benefits, which provide, inter alia, medical benefit like that of health insurance during the first five years of invalidity.

cost of this additional "cure" for two years or more in appropriate sanatoria or preventoria. How much will be contributed in the case of each of the latter extended benefits is left to the individual *tarif* funds and to agreements which the funds or regional unions of funds make with the invalidity insurance organizations. It is recognized that preventive care is to the financial interest of the invalidity funds, and division of the costs of such care between the two types of insurance funds is contemplated.

Many of the larger health insurance funds and the unions of funds have taken advantage of the law's specific authorization to acquire, as well as to contract with, convalescent homes, open-air preventoria, and vacation camps (*colonies de vacance*) for children. The cost to the insured at these institutions is usually only 3 or 4 francs a day, the fund meeting the remaining expense. These special "preventive" benefits have been directed chiefly at the tuberculosis problem, which is a very serious one in France.

Medical Assistance Cases

In the case of the insured who are unable to pay their share of medical benefit costs and consequently are eligible for aid from the Medical Assistance authorities, the arrangements are as follows: The authorities pay the doctors directly at the fund's schedule rates after receiving from the fund the usual "reimbursement," i.e. 80 per cent of the rate set for the service rendered. This leaves the *ticket modérateur* share to the authorities. Eligibility for Medical Assistance aid is made dependent upon the insured person's observance of all the formalities and the rules imposed by his insurance fund in regard to medical benefit.

Additional Benefits

Additional benefits may be granted, with the permission and approval of the Minister of Labor, when a fund has sufficient surplus to its credit. Until the latter half of 1935,[42] the first surplus, except when used for grants-in-aid to preventive institutions, had to be devoted to the granting of nursing allowances and "milk bonuses" to the non-working wives of insured workers. The lion's share of surplus funds spent upon extra benefits has therefore of necessity been devoted to this extra maternity benefit.

The amount of expenditures on additional benefits has gone down each successive year, as a natural corollary of the diminishing surplus of the funds. The decrease results from the fact that the cost of ordinary benefits has mounted with each year of the act's functioning, while, because of the severe depression which has persisted in the '30's, the total received for health insurance from paid-in contributions has gone steadily down from 1,406,000,000 francs in 1932 to 1,188,000,000 francs in 1935. In 1935 the expenditures for the first time exceeded the amount of the contributions paid in. The increase in benefit expenditures is caused exclusively by medical benefit costs. Cash benefits have been held down by the depression wage levels.

Below is shown, in comparative tables,[43] (1) the distribution of expenditures on the various services embraced in medical benefit for each of the first four years of French health insurance, (2) the distribution of medical benefit costs in 1935 among the three types of beneficiary (insured, spouse of insured, dependent children), and (3) the distribution of expenditures on additional benefits in 1935.

The most conspicuous facts in the comparative tables[44] are that (1) the smallest increase over the years is shown in reimbursement of *medical* fees paid to individual practitioners. This was only 22 per cent more in 1935 than that in 1930-

[42] The *décret-loi* of October 28, 1935, altered this. See sec. 30 of this enactment.

[43] The four tables were derived from France, *Journal Officiel*, 1937, Annexe, p. 197 and 1938, Annexe, pp. 1,594, 1,595, 1,597, 1,603. As only the hundreds-of-thousands digit is shown, this is a rough comparison.

[44] As only the hundreds-of-thousands digit is shown, they present only a rough comparison as stated above.

DISTRIBUTION OF MEDICAL BENEFIT COSTS
1930-1935
(in millions of francs and dollars)

	Oct. 1930 to Dec. 31, 1931	*1932*	*1933*	*1934**	*1935**
Medical fees	127.6 ($ 8.4)	130.7 ($ 8.6)	139.1 ($ 9.1)	141.9 ($ 9.3)	156.8 ($10.3)
Surgical fees	31.8 ($ 2.1)	37.5 ($ 2.4)	40.6 ($ 2.6)	50.5 ($ 3.3)	55.6 ($ 3.6)
Drugs	143.3 ($ 9.4)	159.8 ($10.5)	170.3 ($11.2)	188.6 ($12.4)	225.9 ($14.9)
Dental fees	16.5 ($ 1.0)	24.9 ($ 1.6)	29.1 ($ 1.9)	35.2 ($ 2.3)	40.5 ($ 2.6)
Hospitals	56.6 ($ 3.7)	99.5 ($ 6.5)	104.7 ($ 6.9)	110.3 ($ 7.2)	114.6 ($ 7.5)
Dispensaries	5.3 ($ .3)	10.2 ($ .6)	14.3 ($ .9)	15.1 ($ .9)	16.0 ($ 1.0)
Reimbursement to Medical Assistance authorities	4.9 ($ .3)	21.4 ($ 1.4)	27.0 ($ 1.7)	33.5 ($ 2.2)	34.6 ($ 2.2)

*Lacks figure for a few unimportant funds. Totals correspond to 98 per cent.

DISTRIBUTION OF MATERNITY BENEFIT COSTS
1930-1935
(in millions of francs and dollars)

	Oct. 1930 to Dec. 31, 1931	*1932*	*1933*	*1934**	*1935**
Medical fees	37.9 ($ 2.5)	35.5 ($ 2.3)	33.8 ($ 2.2)	34.5 ($ 2.2)	33.4 ($ 2.2)
Surgical fees	3.9 ($ .2)	3.5 ($ .2)	3.3 ($ .2)	3.6 ($ .2)	3.4 ($ .2)
Drugs	4.7 ($ .3)	5.0 ($ .3)	5.1 ($ .3)	5.5 ($ .3)	5.4 ($ .3)
Hospitals	4.8 ($ .3)	9.3 ($ .6)	8.9 ($ .5)	10.8 ($ .7)	11.0 ($ .7)
Dispensaries	.6 ($ .04)	.8 ($ .05)	1.0 ($ .06)	.9 ($ .05)	.8 ($ .05)
Reimbursement to Medical Assistance authorities	.4 ($ .03)	1.4 ($ .09)	1.8 ($ .1)	1.9 ($ .1)	2.4 ($ .1)

1931, in contrast, for instance, to a 57 per cent increase in drug benefit costs, a 74 per cent increase in surgical fee reimbursements, over 100 per cent increase in contributions to hospitalization costs, 150 per cent increase in dental costs, and over 300 per cent increase in the amount paid to dispensaries. (2) Almost as much was spent on drug benefit in 1935 as was paid toward medical and surgical fees and to dispensaries.

(NON-AGRICULTURALS)
DISTRIBUTION OF COSTS OF MEDICAL BENEFIT TO INSURED, HIS SPOUSE AND HIS CHILDREN
January 1, 1935, to March 31, 1936†
(in millions of francs and dollars)

	Insured	*Spouse*	*Children*	*Total*
Medical fees (office)	71.4 ($ 4.7)	14.4 ($ .9)	16.0 ($ .1)	101.9 ($ 6.7)
Medical fees (house calls)	49.9 ($ 3.2)	16.5 ($ 1.0)	27. 6($ 1.8)	94.1 ($ 6.2)
Dispensaries	12.2 ($ .8)	2.7 ($ .1)	5.0 ($ .3)	20.0 ($ 1.3)
Surgical fees	42.9 ($ 2.8)	13.2 ($ .8)	13.3 ($ .8)	69.5 ($ 4.5)
Dental fees	38.7 ($ 2.5)	9.7 ($ .6)	2.1 ($ .1)	50.7 ($ 3.3)
Drugs	172.5 ($11.3)	52.4 ($ 3.4)	57.4 ($ 3.7)	282.4 ($18.6)
Hospitals	78.5 ($ 5.1)	15.8 ($ 1.0)	22.0 ($ 1.4)	116.4 ($ 7.6)
Sanatoria	11.5 ($ .7)	.8 ($ .05)	14.4 ($ .9)	26.9 ($ 1.7)
Reimbursement to Medical Assistance authorities	22.9 ($ 1.5)	6.2 ($ .4)	14.7 ($ .9)	43.2 ($ 2.8)

†699 funds.

EXPENDITURES OF FUNDS ADDITIONAL TO REQUIRED BENEFITS
1935 and 1936

	1935	*1936*
Grants to		
Hospitals and dispensaries	6,271,699 ($ 413,932)	2,785,316 ($ 183,830)
Convalescent and rest homes	2,601,911 ($ 171,726)	703,419 ($ 46,425)
Vacation colonies	4,664,156 ($ 307,834)	2,191,841 ($ 144,661)
Maternal and child welfare projects	1,902,778 ($ 125,583)	2,395,905 ($ 158,129)
Tuberculosis Sanitoria and Preventoria	4,553,383 ($ 300,523)	3,435,690 ($ 226,755)
Anti-Cancer organizations	3,730 ($ 246)	17,640 ($ 1,164)
Anti-Venereal organizations	167,606 ($ 11,061)	261,626 ($ 17,267)
General social welfare organizations	3,054,954 ($ 201,626)	1,083,880 ($ 71,536)
Additional Benefits		
Extra sickness benefits	373,663 ($ 24,661)	2,094,772 ($ 138,204)
Extra maternity benefits	24,414,842 ($1,611,379)	20,387,655 ($1,345,585)
	48,008,722 ($3,168,571)	34,357,944 ($2,333,556)

(3) More than 34,000,000 francs ($2,200,000) was paid for medical benefit to insured persons who could not pay their *ticket modérateur*, and who in consequence were provided for through the Public Assistance authorities.

XXI

THE "TICKET MODÉRATEUR"

AS has been stated in discussing the *tarifs de responsabilité*, the 20 per cent share of the *tarif* fees for medical service as to which the patient is not reimbursed is called the *ticket modérateur*, translated roughly as "controlling fee." Like the "controlling fee" in Denmark,[1] this *ticket* is explained as a means of restraint upon the insured who might be inclined to burden his insurance organization with the cost of unnecessary demands for medical attention and drugs. However, the similarity between the Danish controlling fee and the French *ticket modérateur* is very limited in relation to medical bills. In Denmark the insured pays nothing to the doctor directly. The doctor is paid by the society, and the small amount of the control fee which the insured pays to the society is the total amount that the insured is permitted to pay for his insured medical care. In France, in contrast, the insured is billed directly by the doctor and must pay him any amount which the latter deems suitable. The French *ticket modérateur*, which constitutes 20 per cent of the society's *tarif* fee, may be, and often is, only a small fraction of the cost of medical service that the patient must actually absorb personally.

For example, the Parisian worker sometimes must pay 50 francs[2] for a single office consultation with his family doctor. The *tarif de responsabilité* fee for such service, however, is only 15 francs, of which 20 per cent or 3 francs is the *ticket*

[1] It will be remembered that Copenhagen societies charged for a "controlling fee" when the doctor was called on Sundays or holidays and that rural societies, which paid their doctors on a fee basis, imposed this charge quite generally. See pp. 111-12 *supra*.

[2] This is as of the summer of 1936. Both medical charges and reimbursement rates now are altered but the principle illustrated, while less exaggerated than before, remains.

modérateur. In other words, the insured is entitled through his medical benefit to reimbursement of 12 francs from his society toward the cost of this consultation. He must himself bear the burden not only of the 3 francs (20 per cent of 15) represented by the *ticket*, but also the 35 francs which is the amount of the doctor's charge over and above the fee in the *tarif*. The doctors expect in all cases to charge the insured patient a substantial amount over and above the *tarif's* fee, as is indicated by the Medical Association's schedule of minimum fees, which the doctors pledge their organization they will not undercut. These minimum fees in the approved schedules published by the district medical associations are

FEES IN TARIF DE RESPONSABILITÉ
Union of Insurance Societies in Paris and Environs

	Minimum fees Medical Assn. schedule	Urban		Rural		Total to be paid by patient personally	
		Tarif fee	Actual Reimbursement (fee minus 20%)	*Tarif* fee	Actual Reimbursement (fee minus 20%)	Urban	Rural
Office visit	25 fr.	15 fr.	12 fr.	12 fr.	9 fr. 60	13 fr.	15 fr. 40
House calls 9 a.m.—7 p.m. (Weekdays)	30 fr.	15 fr.	12 fr. +2 fr. mileage	12 fr.	9 fr. 60 plus mileage†	18 fr.	20 fr. 40
House calls 7 a.m.—9 a.m. 7 p.m.—10 p.m. (or emergency "at once"—weekdays)	40 fr.	15 fr.	12 fr. + 2 fr. mileage	12 fr.	9 fr. 60 plus mileage†	28 fr.	30 fr. 40
House calls 7 a.m.—10 p.m. (Sundays)	60 fr.	22 fr. 50*	18 fr. +2 fr. mileage	18 fr.	14 fr. 40 plus mileage†	42 fr.	45 fr. 60
House calls 10 p.m.—7 a.m.	80 fr.	30 fr.	24 fr. +2 fr. mileage	24 fr.	21 fr. 20 plus mileage†	56 fr.	58 fr. 80

*Paid only if called especially to attend on Sunday. If in the course of a case which the doctor is treating he makes a Sunday visit, only 15 fr. (99 cents) is paid.

†75 centimes per kilometer in ordinary non-mountainous areas, 1 fr. 25 in mountainous areas.

all markedly higher than the fees in the corresponding *tarif* of the district insurance organizations.

The differential is illustrated by the schedule[3] effective in July 1936 in the Seine region centering in Paris.

Assuming that not more than the minimum of the medical association schedule is charged (which apparently is often not the case), the worker is insured for only 60 per cent of his necessary expenditure for consulting his doctor at the office.

The financial burden of the worker increases with the seriousness of and the unfortunate timing of his illness. Should he require a doctor's services on Sunday, for example, it will cost him 60 francs, of which he can recover only 18 francs from his insurance fund. The 42 francs which he must meet himself is the equivalent of one day's wage of a semi-skilled laborer in Paris. Should he have so unfortunate an experience as to require a doctor's services during the night, his insurance assistance will be proportionately smaller. At the minimum schedule rate, he must pay 80 francs. His reimbursement from the insurance society will be only 24 francs ($2.00). He must himself find about 50 francs. His insurance protection in the last two cases has shrunk to less than one-third of his risk.

The rural worker in this region, in the face of the medical association schedule, is obviously far less completely insured than is his Parisian brother.

A similar disparity shows itself between the reimbursement fees paid by the *tarif de responsabilité* for maternity medical benefit and the minima of the medical association for obstetrical services. The latter are two to three times the reimbursement rates.[4]

[3] The schedule is based upon the rates effective under the 1935 collective agreement. See *Year Book of the Medical Association of the Seine*, 1935, pp. 125*ff*.; also, *ibid*., pp. 104-11. (The reinsurance tarif issued by the central authorities in November 1934 permitted higher rates. See pp. 176-7 *supra*.)

[4] In the spring of 1937 (*Journal Officiel*, 1937, p. 13,246), the first *Tarif de Réas-*

The *ticket modérateur* is theoretically a small charge upon the patient, to make sure that he will not demand unnecessary medical service. But the real "*ticket modérateur*," in the sense of the patient's share in the cost, is not the charge represented by the amount of the *ticket* but is all too often an amount that is more than equal to, and sometimes more than double and treble, the sum which the worker can recover from his insurance organization.

In brief, a very substantial part of even the general medical attention received by insured families in France is not on a health insurance basis in any sense of the word, but is still on a purely private-practice-fee basis.

This has been described by a distinguished French commentator, Étienne Antonelli,[5] in commenting upon the 1930 amendments which introduced the *tarif* provisions, as a *double* control fee unjustified in principle and unfortunate in results: "The law of 1928 provided participation of the insured in medical costs only for the purpose of checking abuse of the insurance privilege. . . . The present arrangement is really a scandalous and abusive interpretation of the *ticket modérateur* principle. . . . When the *caisse* sets up a reimbursement schedule which is less than the medical bill, that involves a primary *ticket modérateur* in itself. Why, then, apply a second *ticket modérateur* based upon the reimbursement schedule?"

surance covering maternity benefits under the revised act of October 1935 was issued as follows:

	Cities with more than 200,000 population	*Elsewhere*
Normal birth	425	325
Instrument birth	425 + 25% to 50%	325 + 25% to 50%
Twins (normal birth)	575	425
Twins (instrument birth)	575 + 25% to 50%	425 + 25% to 50%

[5] Étienne Antonelli (official reporter on the Social Insurance Law of 1930 for the French Chamber of Deputies), "*Le Droit des Assurances Sociales*," Tome I, Part II, Chap. VI, No. 6, published summary of a Course on Social Insurance and Social Welfare given under the auspices of the French Minister of Labor for the *Conservatoire des Arts et Métiers.*

The theory of the French health insurance scheme is that most of the insured are able to pay the excess of the cost of treatment over the part paid as medical benefit by the insurance fund, and that the Public Assistance authorities promptly and adequately supplement the insurance funds in case of workers who cannot finance this excess. The practical facts are often far removed from the theory. This is of special significance in the cases which call for prompt treatment in sanatoria or preventoria. This is a fact frequently lamented by the medical specialists who direct the tubercular-cure placements for the insured population. Time is, of course, of the essence in such placements, since successful combating of tuberculosis is linked with prompt provision of necessary care in the incipient stages of the disease. Yet in cases found to need sanatorium care, there is prompt placement as a rule only in the cases in which the insured can himself pay the part of the cost of treatment which is not met by medical benefit.

This situation is well illustrated by a sampling study of placements for "cure" of tubercular insured persons sent from one of the Paris hospitals during 1934-1936 (*Hôpital Salpêtrière*).[6] Only 20 per cent of the adults whose condition required such cure were placed directly through the health insurance fund. The other 80 per cent could not raise the money to meet the supplementary costs and had to be placed through one or another of the Public Assistance or Social Hygiene agencies. All of the fortunate 20 per cent who could pay their way were placed within a month. None of the assisted cases were placed until the second month. Approximately one-fourth of such cases failed to reach a sanatorium within three months, and 7 per cent were not institutionalized for more than five months after their fund's Department of Cures certified their immediate need of such treatment.

[6] For complete details see *Lenteurs et Difficultés des Placements en Station de Cure des Tuberculeux Adultes et Enfants*, par Étienne Bernard et G. Dreyfeus, in *Revue de la Tuberculose*, July 1936, pp. 856-72.

A similar condition prevailed in the children's division. Placements when the insured paid his *ticket*, were made after an average delay of only three weeks; whereas the minimum delay in the other cases was two months and a third of these children did not reach an institution until more than five months after it was judged necessary.

Both the French administrative authorities and the French doctors recognize that the actual *ticket modérateur* situation is an unfortunate one. They are in agreement that identity between the amounts set in the *tarifs de responsabilité* of the funds and the actual charges of the medical practitioners in treating the insured families would be a desirable situation. There, however, agreement usually ends. The medical associations, for the most part,[7] maintain that the situation should be adjusted exclusively by an increase in the *tarif* rates. The authorities, on the other hand, insist that the medical associations should (1) steadily lower their minimum fee schedules until they approximate the *tarif* rates, and (2) adhere to these minima when treating the insured.

This insistence of the authorities took rather threatening form in at least one instance: In a circular announcing an increase in the reinsurance schedule in the metropolitan areas, the Minister of Labor made the following statement in the fall of 1934:[8]

"Thanks to the means at the disposal of the funds, the latter can, by the interplay of reciprocal concessions, move toward the ideal solution of identity between their *tarifs de responsabilité* and the *tarif syndical* [i.e. the minimum fee schedule of the Medical Association]. . . . My representative sitting with the departmental commission will take an appeal to the Social Insurance Council . . . in the case of approval of any *tarifs* which raise the previous *tarif* rates

[7] This is not the attitude of all medical men in France. See section on pp. 248*ff.* for discussion.

[8] Adrien Marquet, *Circulaire*, issued August 8, 1934. *Journal Officiel*, 1934, p. 859.

without the corresponding concession by the doctors of lowering their association minimum fee schedule. I may add that, if lowered schedules are not in fact adopted by the majority of the medical profession, the annulling of the collective agreements in conformity with Article IX of the model agreement published in the August 8, 1930, *Journal Officiel* will have to be considered." Such annulment would, of course, reduce the amount of reimbursement to the small flat-rate fee for all medical services which is operative in the absence of collective agreements.

As the provision in the model agreement referred to in the ministerial circular states that agreements may be declared null and void in the case of serious shortcoming (*manquement*) established by the departmental social insurance authorities, it may be doubted that the ministerial threat could have been carried out with legal propriety. The agreements, however, while running from year to year with automatic renewal, can be terminated any year on notice. Therefore, with the collaboration of the insurance funds and the social insurance authorities, renewal of agreements and approval of agreements could be prevented. This would of course bring pressure upon the medical group.

This circular caused a furor among the organized medical men, who felt that it was a coercive attempt in violation of the fundamental principles of the French Medical Charter[9] to tie the medical associations' fee schedules to that of the *tarif de responsabilité*. The profession, to quote from the French *Medical Journal's* account of this episode, "refused to accept this ultimatum." The administrative council of the Association[10] took "the steps necessary to the obtaining of a modification or an official interpretation that in practical effect was a modification of the offending circular."

[9] See pp. 173-4 *supra*.

[10] The French Medical Association is really a federation of associations, *Grande Confédération des Syndicats Médicaux de la France*.

As a result the Minister of Labor wrote an official and conciliatory letter[11] to the Medical Association in which he stated, inter alia: "I think I should assure you that there could be no question of entertaining any idea that there is legal and obligatory relationship between the *tarifs de responsabilité* and the medical associations' minimum fee schedules. There can be no confusion on this score. A simple reading of the text indicates that these schedules are independent of each other. My circular should be interpreted as the expression of a desire to see the medical bodies contribute to the lowering of the cost of living by certain sacrifices through which the lower paid insured groups, as is equitable, would be the first to benefit."

[11] See full acount of this episode and the text of the letter by Dr. Cibrie, General Secretary of the Federation of French Medical Associations, in the *Revue Internationale de Médecine Professionelle et Sociale*, November 1934, pp. 112-14.

XXII

THE GROWTH OF THE DISPENSARY AND THE MUTUALIST "CLINIQUE"

THE gap between medical benefit and the actual costs of the services involved has brought a development which the medical profession has found less easily adjusted than the problem of a Minister of Labor and an offending circular. This development is the growth of "pay" dispensaries and private hospitals (*cliniques*) with out-patient departments. These dispensaries are created and maintained not only by the Red Cross and similar agencies, but also by mutualist groups and certain Popular Front communes.[1] The health insurance act specifically authorizes the funds to own and to maintain, as well as to contract with, clinics or dispensaries in which the insured may receive medical and surgical assistance. Through such institutions an increasing number of insured patients have been seeking medical care in lieu of going to private practitioners.

The reasons for this development are apparently uncontroverted. The financial disadvantages to the insured worker of the French type of medical benefit have been so great as to press inevitably toward arrangements within the scheme to cut down the worker's risk of financial loss through medical and hospital bills. The "pay" clinics, since they employ a full-time or part-time medical staff on salary or other similar remuneration basis, can offer a complete medical and surgical service for fees equal to or less than the amounts scheduled in the *tarif*.

The insured who patronizes such a clinic runs no risk of unknown medical bills. Sometimes, notably in the case of

[1] See discussion, Villejuif dispensary, at pp. 242*ff*.

the mutualist *cliniques*, he pays a small extra premium in return for which his fund absorbs his *ticket modérateur* share. Always, if a mutualist, he has the certainty that his burden will be at most merely the *ticket modérateur* amount, i.e. 20 per cent of his fund's *tarif* fee. Some of the Popular Front communes, moreover, have set up municipal dispensaries which do away with the *ticket modérateur* altogether and provide gratuitous medical service at the dispensary for the whole insured population.

This dispensary development has been bitterly resented by the organized medical profession in France, on the score that it cuts radically into the office practice of private physicians and creates a relation between doctor and patient wholly inimical to the principles of the French Medical Charter. Various local medical associations have continuously conducted a vigorous drive against the physicians who affiliated themselves with the "pay" dispensaries, even to the point of expelling certain distinguished members who refused to sever their salaried connections with offending clinics. Their local efforts proving ineffectual, the regional organizations finally carried their complaints to the National Federation of Medical Associations. This organization, in response, created a "Dispensaires" committee,[2] which, in the spring of 1936, formulated an official program of suppression and control to be followed by the Administrative Council of the Federation in a nationwide war upon the dispensary movement. This program, insofar as it relates to dispensaries giving treatment and not mere diagnosis, rests upon the use of official collective pressure upon the public authorities and group pressure upon dispensary doctors, to achieve the following results:

1. Philanthropic and Municipal Dispensaries. Only gratuitous treatment to be given and only for indigents, in any dispensaries which are either partly or entirely endowed by

[2] For detailed report of this committee, see the July 1936 issue of *Le Médecin de France.*

public or private philanthropic subsidies. (Main offender, the Red Cross.)

2. Mutualist Dispensaries. Only gratuitous treatment and only for bona fide insured members of the parent mutualist organization.

3. All dispensaries taking pay patients to be treated as ordinary *commercial* ventures to be frowned upon but accepted, provided that (1) they are licensed as medical offices, and (2) their charges are at least equal to the Medical Association's minimum fee schedule.

Among the "philanthropic" dispensaries, the Red Cross was marked as the first object of concerted attack. It was announced further, that collaboration with mutualist dispensaries was prohibited to all Association members until these dispensaries should have adjusted their arrangements to meet Medical Association demands.

Since the medical profession has insisted on absolutely free choice of medical practitioner by the insured patient, it cannot direct its anti-dispensary campaign against the insured's right to patronize any clinic in which he may choose to be treated. To deny the insured the right to go to a clinic, if so desired, would be, of course, a violation of the privilege of free choice in regard to medical services, which the profession itself vehemently and successfully demanded should be written into the law.

The medical arrangements incorporated in the French Act were, as has been explained, a result of the insistence of the medical profession that the insured patient should freely choose his doctor and pay him whatever the latter demanded, without any regulation and control of the physician's charges or his conduct of the case. This "freedom" achieved by the French profession is in marked contrast to the regulated arrangements under which the British and Danish doctors serve insured patients. The French health insurance arrangements were chosen by the profession in the interest, it was declared, of conserving to the patient and to the medical profession the traditional doctor-patient relationships. Paradoxically, however, it is only in France, and not in Denmark and Britain, that the insured workers in growing numbers receive their medical care from institutions which employ their medical staff on terms of salary or other fixed remuneration. In Denmark and in Britain the dispensary as a medium of insured medical care is almost nonexistent. Free choice of individual practitioner and the doctor-patient relationship of private practice tradition have been preserved by the health insurance arrangements of both Denmark and Britain. Encroachment upon this relationship, however, has been promoted by the unregulated "freedom" of the medical profession in the health insurance scheme of France.

XXIII

RESPONSIBILITIES OF THE FRENCH HEALTH INSURANCE DOCTOR

IT has already been explained that in the French health insurance scheme, the doctor who attends insured patients is not required by law to have direct contractual relationship with either the insurance funds or the government. It has also been explained, however, that much more liberal reimbursement is permitted the insured on his medical bills when the medical association enters into collective agreements with the insurance funds; that this more liberal reimbursement is decidedly in the financial interests of the medical profession; and that such collective agreements are almost universal.

The basic provisions of all medical association contracts must follow the official "model agreement" issued by the Ministry of Labor. Its terms were drawn up in the fall of 1930[1] by a committee of the *Conseil Supérieur* in which representatives of the organized profession participated. These terms reiterate the free-choice-of-doctor principle, which is incorporated in the Social Insurance Act itself, and while they provide that the doctor's bill shall be arranged directly between doctor and patient,[2] they impose on the practitioners certain health insurance duties which are essential to the functioning of the scheme.

In consequence, most of the doctors who attend health insurance patients have certain contractual obligations to the insurance fund with respect to their insured clientèle, and these obligations are identical with those of the British and the Danish practitioner in their health insurance practice.

[1] *Journal Officiel*, 1930, pp. 9,161-2.

[2] See pp. 179-80 *supra* for discussion of agreements that attempted to change this.

Thus the doctor in Denmark, like health insurance doctors in all other countries, must certify his patient's incapacity for work, in order that cash benefit may be applied for. Similarly, he must certify the date at which the patient will be fit for work again, for the guidance of the fund in terminating cash benefit. If the attending practitioner wishes consultation or believes that the patient needs a specialist's services or an operation, the doctor must furnish a signed statement to that effect on official letterhead paper. In addition, he must state on the first insurance report slip (*feuille de maladie*) the probable duration of the illness. He is further obliged to aid the fund in its supervisory services by noting on the report slip his medical instructions to the patient relative to exercise, staying in bed, going outdoors, etc.

He is obliged to record accurately, by use of symbols agreed upon, the bill paid (or due) for each item of service he renders, so as to facilitate the reimbursement to which the insured is entitled under medical benefit. If his charges are less than the *tarif* rate for the medical service in question, he must state the amount of the actual bill. If he has served gratuitously, he must record that fact.

He must sign only for bills related to services which he has given individually, and he must use, in furnishing all data for which he is responsible, the official insurance report slips or *feuilles* with which the fund provides the patient.

In prescribing for an insured patient, he is required to write each prescription on a separate letterhead slip, on which must be placed the letters "A.S." ("Assurances Sociales," signifying it is a health insurance case), followed by the insured's social insurance number. If he deems the patient in need of "auxiliary services," such as massage or medicinal baths, he must give him (again on his own letterhead slips) an order explicitly stating the exact number and nature of the requested services. Finally, he must accept the disciplinary machinery of the medical association (the *Conseil de Famille*) as an agency of supervisory control over the performance of his professional duties with respect to insured patients. He must permit the insurance funds the right to be represented in the disciplinary sessions held by that *Conseil.*

The medical associations themselves are bound to use the same nomenclature in their minimum fee schedules as is used by the funds in their *tarifs de responsabilité*. They are pledged to carry out conscientiously their power of disciplinary control over the health insurance doctors and to finance this control.

All collective contracts made by the medical association and the funds must be made binding for the period of a year, with automatic renewal in absence of a month's notice of termination.

The official *décret*[3] which sets up the social insurance administrative framework, extends the binding effect of these collective agreements beyond the membership of the contracting medical associations, by ruling specifically, that "Every doctor who fills out the *feuilles de maladie* for use by an insured patient is deemed to have accepted the collective agreement with the insurance funds which is entered into by the medical association of the region."

It remains to be emphasized that, as before remarked, the completely free choice of doctor which the law allows the patient, requires that some place to be made for the doctor who will not cooperate in any way with the health insurance scheme. This occasional non-organization doctor, by refusing even to use the insurance report slips, evades any responsibility under the collective medical agreement of his district. Reimbursement of his bills is permitted on a *forfait* basis, i.e. a small flat-rate sum for each item of medical service, whatever its character. The insured can obtain this by presenting receipted medical bills to his fund.

The economic disadvantages inherent in this *forfait* arrangement have been discussed in the last section in connec-

[3] *Journal Officiel*, 1936, p. 3,235. (Title II, Article 10, sec. 1 of the decree.)

tion with the *tarifs de responsabilité.* They are sufficiently pronounced to prevent the number of nonconformists from becoming of any importance. In the vast majority of cases, the doctor who renders medical services to the insured families in France, through voluntary acceptance of the responsibilities imposed in medical collective agreements, furnishes his patient with the standardized data which his fund requires on applications for both cash and medical benefit.

In brief, in every respect except that of remuneration, the French health insurance doctor by a roundabout route arrives at the same regulated position of responsibility to the health insurance scheme as that occupied by the health insurance doctor in other countries. Even as to remuneration he is not entirely unaffected by official regulations. Some of his remuneration comes from the fund, and one of the factors which affects the amount which he can hope to collect from his patient is the *tarif* of the fund.

Attitude Toward Certification Responsibilities

The French health insurance doctor is, as we have seen, for all practical purposes in the same position in regard to the certification duties as are the health insurance doctors of other countries. The duty seems to be accepted without question by the French profession, as in the case of British and Danish health insurance practitioners, as the normal responsibility of an attending physician. The patient must have rest and food; for these he needs his cash benefit; therefore his medical practitioner "certifies" his incapacity for work. Repeated questioning in various parts of France elicited only surprise that objection to the certification duty should be expected. If the obligation is an unduly annoying one, the doctors say nothing of it, and as the French medical journals are open to discussion of the irritating aspects of health insurance, the lack of comment about certification duties seems fair evidence that no certification "problem" exists.

XXIV

"CONTRÔLE MÉDICAL"

THE doctor's "free" position under the French health insurance scheme and the fact that medical benefit consists in cash "reimbursements" to the insured have promoted suspicion on the part of the funds of collusion between doctor and patient for the presentation of fictitious bills and unjustified cash-benefit disability claims. This suspicion has been aggravated by the lack of direct disciplinary control over the doctor by governmental agencies. The added circumstance that the French system contemplates payment on a fee basis for each medical service has created almost frantic concern on the part of the health insurance funds lest the costs of medical benefit become excessive through unnecessary medical visitation and prescription. To protect their exchequers from imposition, most of the larger funds, soon after the institution of health insurance, organized a *contrôle médical* service, i.e. an administrative organization for detecting fraudulent claims of medical and cash benefit. In the larger funds this organization from the beginning has included doctors, hospital *infirmières* (trained nurses), and visiting *infirmières* (combined district nurse-medical social service workers), as well as lay inspectors or "visitors."

Since the revisory administrative regulations in 1936, every fund has been required either to include in its own supervisory control staff at least one medical officer, or else to contract out such *contrôle* to a medically staffed agency set up by another fund or by a union of funds. This same regulatory *décret*, as remarked earlier, requires an authorization from a medical *contrôle* officer in case a specialist's services, a consultation of doctors, or auxiliary services are desired by the insured. Refusal of such authorization puts the insured in the position he is in when refused any other phase of benefit on the score that his physical condition does not justify it. In brief, he must either accept the fund's decision and forfeit reimbursement, or else call for an appeal to the technical commission of three doctors previously referred to.[1]

The professional supervisory staff in its pursuit of fraudulent claims or in passing upon requests for special services, interviews and examines insured patients both in their homes and in the fund's headquarters. The *contrôle* also performs another quite different function, i.e. the ferreting out of cases of incipient invalidity. Health insurance funds are authorized to watch for such cases and to direct them to proper preventive treatment with a view to cutting down invalidity claims. This preventive function is not a mere privilege. It is a duty imposed by the administrative *décret*[2] as necessary cooperation with the funds which administer the invalidity obligations of the social insurance law. Since the first six months of illness are the responsibility of the health insurance organizations, ignoring of incipient invalidity by those funds as beyond their province would seriously increase the ultimate financial obligations of the social insurance institutions. Specific obligation is imposed to notify the appropriate invalidity insurance fund of every case which lasts more than two months and suggests possible permanent results.

The combined functions of discovering unjustified claims for benefit and of directing preventive treatment have caused, in the case of some of the larger insurance funds, the development of *contrôle* organizations which are real diagnostic centers. Here staffs of diversified specialists equipped with

[1] See p. 190 *supra.*

[2] Section 37 of the 1930 *décret*, and Article 28 of the 1936 *décret*. (*Journal Officiel*, 1936, p. 3,237.)

laboratory and X-ray facilities check the physical condition of a large percentage of the insured families, survey the treatment they are receiving at the hands of their chosen practitioners, and plan specific alternative preventive therapy when it seems advisable.

The Grande Caisse Interdépartementale at Paris, which is France's largest fund, insuring more than one-sixth of those covered by the French insurance law, in 1936 had a *contrôle* medical staff of 83. Of this medical staff, 48 were on a full-time and 35 on a part-time basis. The staff comprises physicians in every field of specialization, including men who do nothing but read and interpret X-ray plates in collaboration with specialists in the particular area of the body affected. The *contrôle* headquarters has its own bacteriological and X-ray laboratories. The work of the medical staff is supplemented and implemented by 97 certified *infirmières* of two types—a few trained hospital nurses and the rest district nurses trained to do medical social service work.[3] The latter do full-time visiting in the homes of insured workers, who either apply for cash benefits or make demands on the insurance funds for reimbursement of medical bills. Even this small army of doctors and nurses does not perform all the *contrôle* work for this fund. Supervision of maternity benefits is contracted out to the Public Assistance authorities which, in collaboration with the Social Hygiene Bureau, organize both prenatal care and a maternal and infant welfare service.

Each of the visiting nurses of the Grande Caisse averages ten visits a day and often makes more than three hundred such visits in a month. This corps of workers acts as the first line of defense for the *contrôle*. In case the *infirmière* believes that a case demands a medical check-up, whether on therapeutic grounds or because of suspected fraud, the record card is marked with a red cross. This results in a summons of

[3] In France each of the two categories of *infirmière* must possess a government certificate, granted after a prescribed training.

the patient to the fund's supervisory headquarters for questioning and possible medical examination, or, if the patient is bedridden, in a house visit by one of the doctors on the fund's staff.

In the first six months of 1936, nearly 98,000 insured persons were questioned by the *contrôle* medical staff and more than 95,000 were given medical examinations.[4] This is an average of 32,000 interviews and examinations per month, which is on a scale of 384,000 such check-ups a year for a fund carrying about 1,600,000 insured and their dependents. The magnitude of this supervision may be suggested by recalling the work of the British Medical Referee staff, which in the year 1935,[5] in performing its supervisory checking service for 15,500,000 insured, questioned and medically examined about 230,000 men and women. In brief, the medical *contrôle* of this one French fund checked, by either interview or examination, at a rate of 50 per cent more cases, and actually examined medically at a rate of more than four-fifths as many cases as the British Referee service in functioning for all of England.[6]

Reports of health insurance organizations indicate that they are becoming convinced by experience of the importance of expanding their *contrôle* work. This is illustrated by the experience of the largest Parisian mutualist organization, the Fédération Mutualiste de la Seine,[7] which, because it owns and conducts a clinic and dispensary (including hospital beds for short stays in connection with surgical work) through which many members receive their medical service, has a

[4] Information furnished by the chief statistician of the Grande Caisse, M. Duval, in August 1936.

[5] See pp. 55-7 *supra*.

[6] Dependents are not insured in England. The number of dependent wives and children of those insured in the Grande Caisse is not recorded by the Caisse, so exact figures of those eligible for medical benefit are not available. The estimated number suggested by M. Duval in charge of the Statistical Department of the Caisse is: Wives 15 per cent additional, children 25 per cent additional to the insured who numbered 1,600,000.

[7] With an average membership of about 430,000 in 1933; 434,000, January 1935; 447,528, October 1935.

control problem of smaller proportions than would otherwise be the case. The four *contrôle* physicians of this fund in 1933 examined and checked at their headquarters or in the home of the patient more than 17,000 applicants for health insurance benefits. Feeling the need of more effective supervision, the number of physicians was increased in 1934 from four to eleven and the number of examinations made by the doctors in their control work increased to more than 28,000 and to double that number in 1935. During this period the number of medical *contrôle* centers in the outlying regions also was doubled so as to make more effective the check-up of insurance claims in the suburbs. As a result, 6,744 medical examinations were made in these branch centers in 1934 and more than 12,000 in 1935, in contrast to 689 in 1933.

In the industrial city of Lyons, the departmental fund, La Caisse Départementale d'Assurances Sociales du Rhône, which in 1935 insured about 160,000 workers and their dependent families, maintains a *contrôle médical*, which by 1935 was undertaking to check, within twenty-four hours of application, every case in which cash benefit was requested. Any case considered suspicious by the visiting inspector is reviewed by medical examination at the fund's headquarters if the case is ambulatory, or at the home of the insured if it is claimed not to be so. More than three times as much was spent in 1935[8] on this supervisory service as had been expended for the purpose in 1931. The *contrôle médical* was equipped with complete X-ray equipment and comprised, in addition to four full-time physicians, a corps of part-time tuberculosis specialists and specialists qualified in the digestive and genito-urinary fields. In the year 1935 practically all insured members disabled by illness for more than eight days received actual medical examinations, which averaged between eighty and ninety a day.

[8] 133,296 francs in 1930-1931, 430,851 for 1935. From *Rapport sur le Fonctionnement de la Caisse Départementale d'Assurances Sociales du Rhône pendant l'Année*, 1935, p. 4.

In the city of Nancy,[9] *contrôle médical* is contracted out by the funds to the local medical association. The funds pay 1.3 francs per annum for each insured person for this service. In handling such work for the departmental fund of the Nancy area, which administers the health insurance benefits of 50,000 insured families, the staff of the medical association checks about 1,000 cases a month. Complete medical examinations are made in all questionable cases. Laboratory analyses are included in the diagnosis, which is communicated to the attending physician.

The funds apparently are satisfied with this "farmed-out" *contrôle*, and they do not have the friction with the medical association from which some funds with active supervisory staffs suffer. The Nancy funds believe that the cost of their benefits is at least not more (perhaps less) than it would be if they undertook their own checking and supervision. This situation undoubtedly reflects the Nancy's famous co-operative alliance among the medical association, the social insurance funds, the social hygiene authorities, and other welfare organizations discussed in Chapter XXVI below.

The requirements of the 1936 decree have prompted the small funds in all parts of France to group themselves into "unions" to which they have delegated their *contrôle* responsibilities. They are thus securing the advantages of a supervisory body, well equipped and well staffed, which otherwise would have been beyond their financial reach.

Supervisory control in the rural areas is perforce much less elaborate and more lax than in the large cities. Thus even the departmental fund of the Seine et Marne district, which spreads over approximately 2,280 square miles, has but one medical control officer, although it insures 135,000 workers (more than half the insured of the area). This district is predominantly agricultural, three-fifths of its 534 communes having less than 500 inhabitants. The fund has arrangements

[9] See pp. 236*ff*. for discussion of Nancy's interesting health and hygiene organization under Dr. Parisot.

with the twelve dispensaries of the Bureau of Social Hygiene which permit the control officer to use the dispensaries for examining the insured and which ensure that the twenty-six visiting nurses of the Bureau visit the insured families. The latter service of the Bureau of Hygiene, however, is purely a health service and not for the detection of fraud. Fraud control is exclusively by the physician *contrôleur*, who checks and examines about 1,000 cases a year, mostly in the homes of the insured.

It should be reiterated that all this supervisory control is exercised, of course, directly on the insured. Inevitably, however, since each case involves checking the physical condition of some doctor's patient, it amounts to a review of the attending practitioner's diagnosis. This, quite naturally, has been resented intensely by some doctors who have complained that the supervisory medical staffs were, without excuse, inquiring into the professional handling of cases by the attending practitioner. The latter concededly is, under the French law, wholly beyond the jurisdiction of the funds. Disciplinary scrutiny of the doctor's performance of his health insurance duties, i.e. *contrôle technique*, is entrusted primarily to the medical organizations themselves.[10] Insofar as direct proceedings against the doctor are concerned, it is clear that a fund's administrative staff can do no more in any case than file a complaint with the medical association's disciplinary body, leaving the handling of the complaint to *contrôle technique*.[11]

But the fund's supervisory medical staff can and does advise the fund to hold up benefits when it believes a medical bill unjustified by the patient's condition or deems an incapacity certification by the doctor unfounded. This, through its effect upon the patient's insurance claims, is a judgment, if an indirect one, upon the doctor's conduct of his profession.

[10] See next section for a review of *contrôle technique*.

[11] *Journal Officiel*, 1936, p. 3,235 (Article 16 of the administrative *décret*).

And this, of course, has inevitable repercussions upon his future practice.

The staff physicians of *contrôle médical* have always been forbidden by the regulatory decree either to give any medical treatment or to "intrude in any way into the relations of the patient and the doctor." The original administrative regulations listed,[12] moreover, the matters which are the concern of *contrôle technique*: "All questions of medical treatment, particularly the number of visits made, the accuracy of certification by the doctor, the care with which he prescribes, and in general all questions of the way in which medical benefit is handled by the practitioner."

Despite this, general complaints by the doctors that *contrôle médical* was encroaching upon the province of *contrôle technique* were so numerous that the 1936 regulations attempted a more specific mandate. The fund's supervisory physician is now forbidden "either to express any diagnostic opinion in the presence of the insured person or to make any comment on the treatment ordered by the attending physician." He is also required, if he disagrees with the attending practitioner about the condition of the latter's patient, to get in touch with the practitioner directly.

The very nature of the administrative obligations of the fund, however, i.e. the elimination of fraudulent or unjustified claims and the detection of incipient invalidity, makes it a practical impossibility to arrest the tendency of *contrôle médical* to develop into a diagnostic service, which, confessedly or not, is checking general medical work. The detection of unjustified claims requires medical examination, and the obligation to prevent invalidity when possible justifies examination, review of treatment, and even replacement of treatment with institutional or other preventive care deemed necessary by the fund. This has made impracticable the promulgation of rules effective to insure separation of those

[12] Article 26, sec. 2, 1930, administrative *décret*, *Journal Officiel*, 1930, p. 8,453 (or see *Bulletin du Ministère du Travail*, p. *154).

cases in which complete diagnosis and investigation of treatment being received is justified from those in which it is not. The trend in the development of *contrôle médical* toward diagnostic review of most cases would not appear to be susceptible of adjustment by "official regulations."

There is evidence that this situation is well understood in organized medical circles. Further, there is encouraging evidence that at least the central offices of the Federated Medical Associations of the Seine (Paris) are pressing a policy of collaborating with, rather than fighting against, the medical *contrôle* staffs, to the end of establishing an *entente cordial* in the interests of the proper functioning of the health insurance act. This policy is well exemplified by the addendum of 1935 to the collective agreement between this Medical Federation and the Paris Union of Health Insurance Funds. This addendum reads, inter alia:[13] "The Federation of Medical Associations and the Union of Funds agree to act *together* to put into practice (1) in *contrôle médical* (without violating preservation of professional secrecy or the normal rules of medical ethics) direct communication of information from the attending physician to the medical director of *contrôle*; (2) to study ways and means of developing an individual health record for each of the insured; (3) to the development of all possible statistics of social and medical interest; (4) to the organization of early diagnosis of disease and its prophylaxis, to be realized by the creation of centers of diagnosis to which doctors will send their patients, to be returned with diagnostic advice for treatment by their own physicians."

The *contrôle* situation in France is again a paradoxical one, as is the trend toward salaried work which was remarked in connection with the dispensary development in France. The French organized medical profession shaped the French health insurance scheme primarily with the determination

[13] See *Year Book of the Medical Association of the Seine* (*Annuaire du Syndicat des Médecins de la Seine*, pp. 109-10).

that the doctor's service in attending his insured patients should be completely independent of any supervision or review other than that undertaken by the organized medical profession itself. Yet the work of the French health insurance doctor is in actual fact scrutinized, checked, and reviewed to a far greater degree than is the case in countries in which the health insurance doctor is, theoretically, far less independent. The checking and review, moreover, is in France done directly by the insurance funds.

XXV

"CONTRÔLE TECHNIQUE"

EXCEPT in the case of hospital doctors, supervisory disciplinary control (*contrôle technique*) of the French health insurance doctor is left, in all districts in which collective agreements are concluded between the funds and the local medical associations, to the medical *Conseils de Famille*, i.e. the disciplinary councils of the organized medical bodies. As has been stated previously, doctors, whether members of the medical association or not, are deemed to subject themselves to this disciplinary control when they sign insurance slips for their patients.

Disciplinary control of public hospital doctors is exercised (1) as provided by the law governing the public institutions in question, and (2) subject to such law, as provided by contracts between the insurance funds and the institutions. In the case of doctors performing services in private hospitals and sanatoria and nursing homes, it must be exercised by one of the medical organizations with which the insurance funds have collective contracts. However, the *décret* setting up the basic health insurance administrative regulations has always forbidden such *contrôle technique* to a medical organization composed exclusively of the physicians attached to the establishment.

Disciplinary jurisdiction over the non-institutional doctors is exercised by a local medical council with a right of appeal to the *Conseil de Famille National*, an adjunct of the official National French Medical Association (*La Grande Confédération des Syndicats Médicaux de la France*).

The usual medical council comprises a body composed of eight regular members and four alternates, chosen by secret ballot at a general meeting of the medical association. The council may proceed against a doctor either on its own initiative or on the basis of a complaint filed with it.

The required procedure is simple. If based upon charges filed (the usual situation), a written complaint must be sent to the president of the medical association in question, who in turn passes it on to the president of the council. The latter designates one of the members of the council as official reporter and convenes the council. The accused doctor is formally notified of the charges against him and of the date set for his hearing. Both the defendant and the complaining party are allowed to be assisted at the trial by either another physician or an attorney.

The following penalties[1] are contemplated by the rules of the *conseils*.

1. Warning without notification of the insurance fund.
2. Warning with notification of the insurance fund.
3. Reprimand.
4. Fine.
5. Temporary suspension of the right to treat persons insured in a given fund.
6. Permanent suspension of the right to treat persons insured in a given fund.
7. Temporary suspension of the right to treat any insured persons.
8. Permanent suspension of the right to treat any insured persons.

In the belief that there was need of an appeal from the *Conseil de Famille National* to a body which was not exclusively medical in composition, a term in the "model medical agreement"[2] promulgated by the Ministry provided

[1] There seems to be no clearing house records of the penalties imposed by district *conseils*. It is doubted by Medical Association officials that Penalties 6 and 8 have been imposed.

[2] See p. 211 *supra*.

that there should be a final appeal to a national arbitration commission. As a result, the Minister of Labor created in 1932 *La Commission Arbitrale du Contrôle Technique.* This commission was composed of three members chosen by the Social Insurance Council to serve for a term of one year. Two were selected from its own membership, one from among its medical members, and the other from among the representatives of the health insurance organizations. The third member had to be a "neutral" or "public" representative, involved with neither the practice of medicine nor the administration of health insurance, and was chosen from the *Conseil d'État* (a semiadministrative, semijudicial body, which gives expert legal counsel to Parliament and the Cabinet and is also the highest tribunal for administrative disputes).

The authority of this Arbitration Commission until 1936 rested upon "voluntary" contract between the funds and the medical associations. Since then, as a result of a new section in the revisory *décret-loi* of October 1935,[3] it has had a legal existence and the official title of *Commission Supérieure de Contrôle.* Its composition is that of its voluntary predecessor: three members, two chosen by the Social Insurance Council from its own membership, the third representative of the public and appointed by the *Conseil d'État.*[4] The law provides that appeal from the *Conseil National* may be taken to the Arbitration Commission by the medical association, the physician penalized, or the interested insurance organization.

This newly created *Commission Supérieure* has a broader jurisdiction than that of the *Commission Arbitrale.* It is not only a court of last resort for disputes appealed from the National Medical Disciplinary Council, but also a court of

[3] Article 8, sec. 5.

[4] The representative of the insurance fund is the president of the National Federation of the Departmental Social Insurance Funds, M. Yung. The representative of the doctors is the permanent grand secretary of the French Medical Association, M. le docteur Cibrie.

first and last resort of *contrôle technique* cases which involve misconduct of doctors over whom the medical councils have no authority. The latter situation exists wherever collective agreement has not been entered into between the insurance funds and the medical association.

When acting as a court of first and last resort, the commission follows the procedural technique of the medical councils, bringing suit against the doctor either on its own initiative or on the complaint of an insurance fund. In these cases it is obligated to ask the advice of the departmental medical association before rendering a decision. If the medical association does not send its opinion within twenty days, the commission proceeds to judgment without it.

In all cases before the commission, whether the case is one of appellate or of original jurisdiction, both sides to the controversy must present their case in writing. Additional oral argument, however, may be dispensed with in appellate cases. If it is to be required, notice must be sent the parties by registered mail within five days of the receipt of the appeal petition. Appearance may be made personally or by attorney. The commission is authorized to obtain evidence from expert witnesses and to undertake any investigations which it deems pertinent.

A doctor accused of misconduct in connection with his practice among the insured population thus has, in the usual situation, three possible hearings. The first hearing is his initial trial before the local medical disciplinary council; the second, in the event of an appeal, a review by the national medical disciplinary council; and the third, a final review by the *Commission Supérieure de Contrôle*. The latter is given authority, in the law giving it legal existence, to administer only three types of punitive sentence: (1) warning, (2) reprimand, and (3) temporary or permanent suspension of the right to attend insured patients. The right to impose a fine has not been included in the authorized penalties.

In addition to the *contrôle technique* just outlined, which the medical organizations and the appellate commission exercise over the doctors rendering medical services to the insured,[5] the French Social Insurance Act has always provided special penalty sections which are enforceable through the courts. These sections cover several forms of medical misconduct. Thus, attempting to "attract or hold a clientèle of insured patients by either threats, bonuses, or promises of rebates on medical fees charged" is made subject to both fine and imprisonment. In addition, it is stipulated that a second offense should entail both the foregoing penalties plus the added punishment of publication of the judgment in the daily papers at the expense of the culprit. Finally, it is provided that doctors who intentionally file false statements with the insurance funds may be excluded from health insurance practice and, if collusion with the insured person is involved, subjected to fine and imprisonment as well. These penalties are, of course, like all other punishments for crime, to be imposed only after due trial and conviction.

It has been argued[6] that from a strictly legal standpoint these penalties are the only penalties actually enforceable against the physician and that the disciplinary effect of the decisions of the "extra-legal" medical organization's *Conseil de Famille* rests in the control which the organizations have in fact over the allegedly offending physician, namely, the right to exclude him from the professional advantages of membership in the medical association. It is pointed out that legal provision for enforcement of a penalty inflicted by a medical council is conspicuously absent from the health insurance act. Ergo, it is alleged that if the penalized physician does not value medical association membership, there is no direct force in the punitive sentence imposed by the medical organization's council. The French medical associations

[5] The dental, pharmaceutical and midwives' organizations are given similar disciplinary authority over their members.

[6] See Antonelli, *op. cit.*, Tome I, Chap. III of Part IV.

admittedly do not have power to control entry into health insurance practice, which is possessed by the Danish medical profession.[7]

Important indirect effectiveness, however, is claimed for the council's judgments of "suspension" through the effect of this suspension upon the insured's medical benefit rights. In making collective contracts with the medical groups, the funds are required by the "model agreement"[8] to include a section pledging themselves to respect the disciplinary penalties imposed by the medical councils. From this fact it follows that the funds bind themselves not to reimburse the insured for medical bills paid to a physician who, in defiance of a medical council's suspension sentence, continues to treat insured patients. Such refusal on the part of the fund, of course, operates to draw away the doctor's clientèle, the members of which do not wish to forfeit their insurance benefits. This indirect control over the physician is counted upon to put teeth into the council's decisions.

The funds in fact keep a suspended list and notify their members of each addition thereto, using, under advice from the Minister of Labor, the following form:[9]

> The fund of————hereby notifies its members that from the following date————to the following date ————no reimbursement will be made in respect to services furnished by Doctor————.

It has been suggested by labor law technicians[10] that the legal propriety of even this indirect pressure upon a suspended doctor (i.e. refusal to reimburse the insured for med-

[7] It will be remembered that Danish health insurance practice is open only to members of the Danish Medical Association. See p. 119 *supra*.

[8] Promulgated by the Ministry. See p. 211 *supra*.

[9] See *Le Médecin Conseil* (a journal published by the *contrôle* doctors), November 1934, pp. 27-9, for discussion of the use of this notice.

[10] For such an interpretation of the law see the analysis of M. Antonelli, professeur à la Faculté de Droit, *Rapporteur de la Loi du 30 Avril* 1930, *à la Chambres des Députés*, in the text of the Social Insurance Lectures given under the supervision of the Minister of Labor in the winter of 1932 and 1933, published in 1933 by the Union des Caisses d'Assurances Sociales, de la Région Parisienne, 70 Rue de Rochechouart, Paris, *Le Droit des Assurances Sociales*, Tome I, Part IV, Chaps. III and VI.

ical bills paid to him) may be questioned as in conflict with the fundamental guaranty of section 4, Article 2, of the Act, which provides without any limitation that "the insured has free choice of physician." Since, however, the law clearly contemplates that the medical associations shall discipline offending health insurance doctors and since it also directs the funds to make collective agreements with the medical associations, this seems an untenable suggestion.

From a practical standpoint, moreover, the mere notification of the insured that the fund will not give reimbursement on medical bills paid to a certain doctor (even assuming it might ultimately be compelled to do so) usually is sufficient to ruin the doctor's health insurance practice. The vast majority of working people are not inclined to enter into a struggle over their rights. Moreover, at least in the case of suspensions administered by the now officially established appellate body, the *Commission Supérieure*, the fund which pays out reimbursement benefits at the instance of a suspended doctor is financially protected, for the law provides:

"Every doctor who disobeys the decisions of the *Commission Supérieure* by treating and prescribing for insured persons when he has been suspended from health insurance practice, must reimburse the fund in the amount of any benefits paid out for such service (including cost of dental or other services) which the fund has been led to pay through his authorization."

There now is, therefore, unquestionable if indirect *legal* force in the suspension sentence of the *Commission Supérieure*. Patients who draw health insurance reimbursement benefits for services rendered by such a suspended doctor do so at ultimate financial cost to the penalized physician.

Thus there has always been *practical* force in *contrôle technique's* suspension sentences and actually in the less severe penalties as well, since they threaten later suspension if the doctor does not mend his ways. It is of interest, therefore, both to summarize briefly the decisions of the appellate commissions in the first years of operation of the French act, 1930-1936, and to sample the work of the medical disciplinary councils.

The *Commission Arbitrale* met in two different years, once in 1932 (in December) and twice in 1933 (in May and December). Its legalized successor, the *Commission Supérieure* convened three times in 1936 (May, June, and October). Altogether twenty-six punitive judgments were handed down, one of which was annulled on rehearing, leaving twenty-five effective sentences. Three were registered the first year in which the commission sat, five the second, and seventeen (besides one which was annulled on rehearing) during the third. Three of the sentences involved merely a "warning," five were "reprimands," and seventeen were temporary suspensions from health insurance practice.

It should be borne in mind in surveying the suspension sentences that the commission, like all appeal bodies, may pass only upon the issues presented to it. It may not, when not so requested, administer heavier penalties than those involved in the record, even though it might feel inclined to do so. Thus permanent suspension has never been at issue before the commission. Of the temporary suspension sentences, the longest suspension period was five years. This was inflicted in penalization of several forms of misconduct, the most flagrant of which was collaborating with a "Medical Institute" of charlatan character, for the patients of which the doctor signed medical reimbursement slips even when he did not perform the medical services. The second longest suspension was for three years, imposed for practising medicine by correspondence. Next in severity was a two-year suspension sentence which was administered for another "Institute" offense. A year's suspension was adjudged in a similar case, i.e. collaborating with a "Medical Institute" of questionable character which manufactured a proprietary medicine which, regardless of the illness involved, was constantly prescribed by the penalized physician.

The remaining thirteen suspensions were for periods of less than a year, six being for six months and the remaining seven for shorter periods.

Except in the "Institute" cases, the penalties have been relatively mild in view of the misconduct proved. This seems especially true when contrasted with the penalties inflicted by the British disciplinary machinery from the first years of operation of the act.[11] Permanent deprivation of the right to health insurance practice was frequently inflicted by the British for deliberate violation of professional obligations similar to that involved in cases which *contrôle technique* of the French Medical Association disciplined far less severely.

A variety of factors contribute to the difference between French and British disciplinary experience. The organization of health insurance medical service in Great Britain is entirely different and the effective legal control over the conduct of the doctors is unquestioned. In addition, the disciplinary machinery from its initial stages, while participated in by the organized medical profession, has not an exclusively professional membership, while in France only the highest appeal body has lay as well as medical members.

It should be added that any professional disciplinary group when dealing with new social obligations tends to be progressively more strict with its members with each year of its functioning. (This is perhaps suggested by the fact that three five-year suspensions were inflicted in 1937, in all instances for fraudulent imposition on health insurance funds.) It may be anticipated that definitely reprehensible conduct, such as falsifying records to collect fees for services never performed, will lead to heavier and heavier punitive sentences with each year of experience.

The number of cases which reached the social insurance appeal body between 1930 and 1936 was just about half the number passed upon by the highest medical association

[11] See p. 74 *supra* for British discussion, especially n. 20.

court, i.e. the *Conseil National*,[12] and the latter number is of course only a small fraction of the cases involving alleged misconduct of health insurance doctors, all of which pass first through the disciplinary machinery set up for the local medical organizations. This fact is illustrated by the following statistics of one of the French departments, the Departement du Nord. In this department,[13] about 800 doctors practise[14] and 825,000 insured persons and their families receive health insurance benefits.

During the first three years of the operation of the health insurance act, the *Conseil de Famille du Nord* held twenty-four sessions and considered 189 complaints against the health insurance doctor. In 153 instances hearings were actually conducted and in 87 of these (56 per cent of the cases) judgment was rendered against the doctor. Twenty-nine of the decisions involved merely a "warning" of the physician; thirty-four more administered an official "reprimand"; in five instances the doctor was subjected to a fine, and in nineteen the doctor was temporarily deprived of the right to insurance practice.[15]

Eight cases were appealed to the national medical disciplinary council. Three of the appealed cases involved censure only. Five were cases of suspension of the right to health insurance practice. All three of the judgments of censure were annulled by the highest medical body (the *Conseil National*), but two of these were restored on second appeal by the *Commission Arbitrale*. On the other hand, only one of the suspen-

[12] Information furnished by the *Commission Supérieure*, January 1938.

[13] The Department is the main French administrative unit. It has, like the American state, neither a rational geographic nor rational population base. The revisory act (1935) put social insurance administration on a regional rather than departmental basis. See p. 171 *supra*.

[14] 757 members of the Medical Association (data furnished by M. le docteur Hilaire, Social Insurance Secretary of the National Association, i.e. La Grande Confédération des Syndicats Médicaux Français).

[15] One fine was nominal (10 francs), three were in the amount of 200 francs, and one was in the amount of 500 francs. See *Le Médecin Conseil*, Vol. I (July 1934), pp. 419-20. Nine suspensions were for one month, six for three months, three for six months. *idem*, p. 420.

sion judgments was annulled by the *Conseil National*, although one of the four suspensions which it affirmed was not executed when carried on second appeal to the *Commission Arbitrale*.

Thus, of the 189 complaints filed and the 153 cases actually conducted and resulting in 87 disciplinary judgments of the local *Conseil de Famille*, only eight, or less than 10 per cent, reached the national medical *contrôle* court and only three were carried to the highest appellate body.

XXVI

EXAMPLES OF SPECIAL LOCAL SCHEMES PROMPTED BY THE SOCIAL INSURANCE LAW

"Coordination" at Nancy

IN the region centering in Nancy, a most interesting complete coordination of medical, social, welfare, social insurance, and public health services has been functioning since inauguration of the social insurance act. This is, of course, not the only instance of coordination of health and social services,[1] but it is the first instance of complete coordination of all the agencies of a community, and still the most outstanding one. Its nucleus was worked out in the 'twenties, and from the beginning it has been administered under the leadership of the internationally distinguished physician and social hygienist, Professor Jacques Parisot,[2] who was its originator and guiding genius.

The present scheme[3] rests on a formal agreement of collaboration which was entered into, in 1930, by the District Medical Association, the Office of Social Hygiene, and the District Federation of Social Insurance Funds. A central coordinating committee was created, which today represents all the public and private medical and welfare agencies in

[1] In many departments of France (conspicuously La Loise, Le Nord, La Girdone), the insurance funds have made agreements of collaboration with various welfare and health agencies. An example of a coordination scheme patterned exactly after the Nancy scheme is that in operation in La Sarthe.

[2] Professor of Hygiene and Social Medicine (Nancy), member of the Comité Permanent d'Hygiène of the League of Nations.

[3] See, for Dr. Parisot's description of this Nancy experiment, Jacques Parisot, "La Collaboration, Base de l'Action Sociale," *Revue d'Hygiène et de Med. Sociales*, April and June 1932, and for a detailed discussion of all its phases; also see M. Chavarot, *Directives Propres à Orienter les Caisses d'Assurances Sociales dans leur Rôle en Médecine Preventive*, Nancy: George Thomas, 1936, Chap. VI.

fields directly or indirectly concerned with prevention of illness and improvement of public health. These include not only the regional Federation of Insurance Organizations (*L'Union des Caisses*), the District Medical Association, and the local Bureau of Social Hygiene (*L'Office d'Hygiène Sociale*), but also the Welfare Bureau of Nancy, the Hospital Council, the administrative officials in charge of the inspectorates of Public Assistance, the Departmental Public Health Authority, the regional Social Insurance Office, and the local government bureau which administers family allowances.

Three principal collective agreements have been entered into by the local federation of social insurance funds. The first is the main tripartite agreement with the Bureau of Social Hygiene and the medical association. The second is a tripartite agreement with the Hospital Council and the medical association, and the third is a bipartite agreement between the health insurance funds and the medical association. This last agreement[4] provides, inter alia, that the medical association shall do the supervisory claims-checking for the funds.

The objectives of the main tripartite agreement have been to promote preventive work in the Meurthe-Moselle area by placing the diagnostic facilities of the Bureau of Social Hygiene gratuitously at the service of the insured families under arrangements that will ensure the use of these diagnostic facilities. The bureau has agreed that its dispensaries will serve merely as diagnostic centers and in general will refrain from giving actual treatment and prescriptions. In return for this, the doctors agree to collaborate with the dispensaries and with the bureau's fine bacteriological laboratory, which provides gratuitous analyses for the doctors' insured patients. In case of certain diseases, moreover, such as venereal disease, cancer, and tuberculosis, it is agreed that the insured

[4] See p. 220 *supra*.

may be *treated* gratuitously as well as given diagnosis at the bureau's dispensaries. The visiting nurses attached to the dispensaries are used as aids to the *contrôle médical* of the local medical association, ensuring obedience of the attending practitioner's order to his insured patients. In addition, these *infirmières* function as public health agencies to give instruction and advice in the general principles of hygiene, and also as specific supervisory aids of the Social Hygiene Service in direct administration of maternity and nursing benefits of the health insurance scheme. The latter benefits under the terms of the agreements of collaboration are put completely in the hands of the Bureau of Social Hygiene.

The health insurance organizations of the region, in addition to paying an annual per capita contribution of 2 francs (13 cents) per insured for the foregoing advantages, have donated constantly increasing annual subsidies[5] to the bureau to make possible an extension of its equipment and services. As a result, between 1930 and 1935 the bureau increased its dispensaries from 20 to 31, added substantially to its medical specialists, and almost doubled the number of its visiting nurses.

The functioning of the Nancy scheme of collaboration may be illustrated by the steps taken in deciding upon applications for special cures. When an insured person, in the opinion of his attending physician, needs curative sanatorium treatment or a preventive rest cure, the attending doctor, in compliance with the legal regulations, notifies the *contrôle médical* service, i.e. the local medical association. The latter, through its *contrôle* service, directs the insured to one of the Social Hygiene Institute's dispensaries for examination. The dispensary then makes a complete examination (with bacteriological analysis, X-ray, and any other diagnostic aids called for) and reports back to the insurance fund concerned. The report, without violating the "professional secrecy"

[5] 379,265 francs, 1931-1932; 639,980 francs in 1934. See M. Chavarot, *op. cit.*, pp. 223*ff*.

principle, which, it will be remembered, is insisted upon by the French medical profession, advises the fund fully on the following points: the urgency of the case, the nature of treatment advised, the place and type of institution best suited to the needs of the case, and the probable duration of stay or rest cure required. The insurance society, however small, is therefore in a position to act on the application with assurance that it has had excellent scientific guidance.

The maternity service organized within the health insurance benefits is in Nancy, as in many other parts of France, one of the most emphasized phases of the health insurance scheme. As elsewhere, the health insurance authorities, conscious of the low birth rate and the very high percentage of infant mortality, have made special efforts to promote intelligent and adequate prenatal and postnatal care, as well as proper attendance at confinement.

The Bureau of Social Hygiene in Nancy, under the terms of the collaboration agreement, has the administering of all maternity benefits. Not a cent is paid by any health insurance fund for obstetrical service, except on the approval of the bureau's Maternity Service Department, which is granted only when specified requirements as to prenatal supervision have been met. To induce the insured herself willingly to take the initiative in obtaining prenatal care, a cash bonus of 50 francs (over and above the regular insurance benefits) is offered each woman who follows the routine prescribed by the institute. This bonus is the equivalent of two days' wages of adult unskilled male labor in Nancy.

This routine calls for registration of the fact of pregnancy with the insurance society as promptly as possible and, in compliance with the law itself, at least before the fourth month. The insurance fund then sets up an individual maternity record booklet (*carnet de maternité*)[6] for her case and

[6] In 1934 in Nancy itself (exclusive of the outlying regions), 713 families that had never had any previous contact with a public health agency came under the supervision of the bureau, through the maternity service.

forwards it to the Maternity Service Department of the Social Hygiene Bureau. The institute then forwards the booklet to her through one of its visiting nurses to whom the supervision of her case is assigned. As the *carnet* indicates, the insured woman must submit to three prenatal examinations, for which she pays nothing whatsoever. The first examination must be made during the first four months of pregnancy; the second during the seventh; and the third examination during the ninth month. The first examination must be a complete medical examination with laboratory analyses and must include the family medical history as well as that of the patient herself. The second and third are supervisory, but must include at least an albumen test. Both the first two examinations must be made by a physician. They may be conducted at one of the dispensaries, at a maternity hospital, or at the office of a physician collaborating with the program. For the third examination the patient is supposed to select the person, either midwife or physician, who is to attend her at confinement or, in lieu of this, the hospital or maternity home at which she expects to have her child delivered.

A complete report of each examination must be filed with the Bureau of Social Hygiene before the service will be paid for. All laboratory tests deemed advisable, at intervals however frequent, are made by the institute's laboratory without charge, and instructions printed on the *carnet* recommend that basic tests be made every eight or ten days during the last few months of pregnancy.

In addition to this maternity service, which is provided both for wives of the workers and also for "working mothers," a supervisory scheme is in operation for working women who are insured in their own right and entitled to nursing benefits. This scheme is directed at bringing the nursing infants under periodic medical supervision.

The insurance organization, on the birth of a child to a working mother, sets up a nursing benefit *carnet* and sends it to the Institute of Social Hygiene. The latter, as in the case of the maternity booklet, in turn forwards it to the insured through one of its public health nurses to whom surveillance of the infant is assigned. The nursing benefit booklet, numbered and registered to the insured, contains general instructions on the care of the infant, including directions for its diet, and specific information about the administrative regulations governing the nursing benefits provided by the health insurance law. It also includes nine detachable coupons for obtaining the nine monthly nursing bonuses. These are redeemable when presented with a proper medical certificate, either in cash, when the child is wholly breast fed, or "in kind," i.e. in allotments of certified milk, when the mother does not nurse her child or when the child is only partly breast fed. Each coupon contains a medical certificate which can be filled out only when the mother brings the infant in for medical examination at one of the bureau's Infant Welfare Consultation Centers or at one of the bureau's dispensaries, and only by the physician in charge. The mother is instructed to present herself for consultation as well, if she finds herself losing weight while nursing her infant. Neither ordinary physicians nor midwives may give medical certificates that will be accepted in lieu of those called for on the nursing benefit coupons.

In addition to public health and medical duties, the dispensaries also function as social service centers of information and advice, and are open for general consultation at fixed hours every week.

As with the larger insurance organizations in Paris, Lyons, and other urban centers, special rest homes and vacation camps have been arranged for in Nancy. This has been done directly by the Bureau of Social Hygiene in collaboration with the medical association and with the insurance funds. Both the homes and the camps are used for preventive care to guard the under-par against invalidity. All the institutions are conducted under medical supervision at the lowest cost[7] which is compatible with their proper functioning.

Municipal Dispensary at Villejuif

One of the interesting developments promoted by health insurance is the communal dispensary, such as that in operation in Villejuif. Villejuif is a Popular Front suburb of Paris with a population of 23,000. Despite the opposition of the local medical association, this small commune in 1932 equipped part of its municipal administration building, the *mairie*, as a health center with bacteriological laboratory, X-ray and ultraviolet ray apparatus, modern diathermy and hydrotherapy equipment, and a gymnasium for corrective exercise and physiotherapy. With a part-time staff of two general physicians and several specialists, including a gynecologist, an obstetrical and "infant care" specialist, and an ear-nose-and-throat man, this institution offers to the general population gratuitous medical consultations in the field of both prenatal and infant care. In addition, to both insured families and to those receiving aid from the Public Assistance authorities, it provides gratuitous treatment and service of all types both medical and surgical such as may be given at the health center itself. To the rest of its population, i.e. the uninsured, this same general service is offered at nominal rates, ranging from 3 francs (20 cents) for general practitioner service to 10 francs (66 cents) for consultation with a specialist or for ultraviolet ray treatment.

The total cost of the dispensary in 1935 was 300,000 francs, of which the major portion, 225,000 francs, came from the health insurance funds and from the "paying" patients of the clinic. The sum of 24,000 francs was allocated to a service which had antedated the dispensary, i.e. the super-

[7] Usually 20 francs per day, four of which must be paid, if possible, by the insured himself as his *ticket modérateur*. If the insured is destitute, the relief agencies or the institute itself (the latter through funds realized by sale of antituberculosis stamps) furnish the amount of the *ticket*.

visory care of infants and nursing mothers. The clinic service had thus added in 1935 only a little over 50,000 francs to the Villejuif budget, rather than the 300,000 francs which its expenses totaled.

The opposition of the local medical association until the present time has stood in the way of adding domiciliary medical service to the dispensary arrangements. The health center staffs are convinced, however, that such domiciliary service, with or without the cooperation of the organized medical profession, will be realized in the near future as a result of popular demand. In support of this expectation, they point out that the patronage of the dispensary has grown constantly since its inception; that the commune has now purchased and equipped a substantial old building of more adequate size and is planning a maternity hospital; that the number of medical services rendered in 1935 was more than double that of 1933 and was averaging more than 200 a day by 1936, i.e. well over 3 per year per communal resident.[8]

The Surgical Mutual Insurance Fund at Lyons

In Lyons the medical association has been promoting a venture that is of definite significance in the light of certain much discussed principles of the French Medical Charter.

As has been previously explained in describing the French scheme's medical benefit, the organized medical profession in France at the time of enactment of the social insurance law registered adamant opposition to three of the basic procedures of other health insurance schemes. These basic arrangements are (1) that medical services are guaranteed to the insured; (2) that the amount of the doctor's bill is regulated by contract between the doctor and the insurance fund or by the governmental authorities; and (3) that the doctor is paid directly by either the insurance fund or some governmental agency and not by the patient. As a result of medical opposi-

[8] Statistics furnished by the dispensary director, September 1936.

tion, it will be remembered, these conventional medical benefit provisions which were included in the original French bill were eliminated and replaced by the "free" medical arrangements which now govern health insurance service in France. It has already been commented upon that, through the development of *contrôle médical* on the one hand, and the pay clinic and dispensary movement on the other, the medical profession has found disturbing inroads being made upon its actual, in contrast to its theoretical, freedom. There has also been reference to the current bitter struggle on the part of the medical profession to check the growth of pay dispensaries.

A recent development in Lyons, which followed on the heels of a local chapter in this struggle, is found in the surgical insurance scheme[9] which the organized medical profession in Lyons is sponsoring as a current experiment. This scheme was promoted after an unsuccessful effort to check the growth of a large popular mutualist pay clinic and hospital service, by condemning and expelling the doctors involved.

The offending service was furnished by a mutualist fund, the *Caisse Mutualité Sociale*, which, after the social insurance law was enacted built a handsome hospital that was not only ultramodern in its equipment but also gay and noninstitutional in its decorative scheme. This *Clinique Mutualiste*, which was a closed, paid-staff institution, proceeded to offer, in addition to hospital care, an elaborate outpatient clinical service,[10] which scheduled services in twelve special fields of medicine as well as in general medicine.

The *clinique* made arrangements with every social insurance fund in the region. Its services were made available to two groups of patients: (1) socially insured workers (and

[9] Such schemes developed independently of the health insurance act or of organized medical cooperation. For an account of the Second Congress of these mutuals held in 1936 at Toulouse, see *Bulletin Officiel du Syndicat des Médecins du Rhône*, July 1936, pp. 443-6.

[10] The clinic also maintains a laboratory service (open five hours daily) and a dental clinic.

families), and (2) non-socially insured persons who are members of mutualist funds (and families).

The *clinique's* hospitalization charges include both medical and surgical care. No additional medical fees on either a daily *forfait* or an inclusive lump-sum basis are charged. It offers very economical and excellent inpatient service and also provides outpatient consultations which cost at most only the legal *ticket modérateur*, i.e. 20 per cent of the regional health insurance *tarif's* reimbursement fee. In addition, the *clinique* makes available to its own members for a small annual premium,[11] complete insurance of hospitalization and outpatient clinic costs. For this same premium it offers to other mutualists (whether or not socially insured) insurance of well over 85 per cent of these costs.

The *clinique* secured on its staff, physicians and surgeons of fine medical reputation, and the number of both inpatients and outpatients which it attracted apparently not only annoyed but also alarmed the organized medical profession in the region. This alarm has resulted in a venture by the medical association which embodies a realistic approach to the situation that is both in contrast to the usual medical association "war on dispensaries" and also hard to reconcile with certain of the traditions of the French Medical Charter. The regional medical association has sponsored and collaborated in a mutual insurance project, which for all practical purposes accepts the very principles which, through the efforts of French organized medicine, were discarded from the French Health Insurance Act.

The *Caisse Mutuelle Chirurgicale Rhôdanienne* offers for small annual premiums (24 to 36 francs a year in case of socially insured members) insurance of the following character: It will "reimburse" its members, according to an established fee schedule, for surgical fees and also for hospitalization costs up to 40 francs a day for the usual period

[11] For socially insured mutualists only 24 francs a year, 36 francs for other socially insured persons.

required by the particular operation. In the event of complications entailing a prolonged hospital stay, it meets half the additional expense. It permits free choice of any surgeon from a list comprising practically all the surgeons of Lyons except those attached to the *Clinique Mutualiste*, and allows free choice of listed private hospitals and nursing homes.

The word "reimburse" in the previous paragraph is placed in quotation marks advisedly. The actual practice in the case of socially insured persons, is to pay the institution directly the difference between the amount paid by the compulsory insurance fund and the hospital charges, and to give to the insured person a check in the amount of a stipulated physician's bill, *the check being made out so that it is endorsable only to the doctor concerned.*

The Surgical Fund does not actually guarantee that no medical bill beyond this reimbursement amount will be presented, but merely stipulates in its prospectus that the district medical association in its agreement with the fund "strongly recommends" that such be the case. Actually, as it is understood apparently by all concerned,[12] by agreement between the medical association and this *Caisse Chirurgicale* (described locally as the medical association's alter ego), the surgeons are bound to present no additional bill either to socially insured patients or to non-insured persons of similar income levels or to other heads of families who have at least three children.[13]

The significance of such a venture as this in Lyons is that by voluntary supplementary insurance, prompted and collaborated in by the organized medical profession, the conventional health insurance principles of payment of the

[12] This information was furnished the author (in interviews in August 1936) by M. le docteur Chalier, president in 1936 of the French Medical Association (Grande Confédération des Syndicats Médicaux de la France) and Docteur Xavier Delore, vice-president of the Caisse Mutuelle Chirurgicale Rhôdanienne. Both are surgeons practising in Lyons and on the list of the Caisse Chirurgicale.

[13] Members of superior income who insure are guaranteed that supplementary bills, if any, will not exceed 50 per cent of the Caisse's reimbursement tarif schedule.

doctor *not* by the patient but by the insurance fund, in an amount determined *not* by the patient and doctor but by the fund and the doctor, are being embraced. This is being done, of course, in the interest of a principle deemed of greater importance to the doctors than those abandoned in the scheme, i.e. the principle of "free participation of all doctors and free choice of doctor by the patient." In inaugurating this venture, the medical association of Lyons undoubtedly has come to a more sympathetic understanding of the British and Danish practitioners' approval of conventional "medical benefit" arrangements than was entertained by French organized medicine in 1930.[14]

[14] It should perhaps be stated that certain of the surgeons in Lyons maintain that even though the check for medical services is endorsable only to the doctor, the giving of the check to the patient, who in turn gives it to the doctor, preserves the charter requirements!

XXVII

ATTITUDE OF THE FRENCH MEDICAL PROFESSION

AFTER more than six years' experience with the French type of medical benefit, the medical profession generally seems satisfied that it is a basically sound arrangement. There is certainly widespread, definite, and even bitter objection to any radical change, especially to the replacement of fee payments by capitation. While an occasional physician expresses a definite preference for the British system of medical remuneration, there is real emotional intensity in the feeling against "capitation" entertained by many practitioners in France. This apparently is related, in part at least, to the historical fact that traditionally medical assistance for the dependent poor was paid on a capitation basis of an admittedly shockingly low level, which prejudiced the organized profession against the capitation institution itself.

The distinguished social hygienist, Dr. Jacques Parisot, with his passionate interest in public health and in preventive medicine, believes that the basic elements of the French scheme in the long run will be productive of "ordered liberty," yielding better results for both the insured and the medical profession than any other organization of medical benefit. He believes, however, that the general practitioner, not the patient, should select the specialist. The *ticket modérateur* doctrine, i.e. that a part of the cost should be uninsured, he finds a sound one, although destructive when the *ticket* which the insured must pay is too large in amount, as he judges it now often is. He has confidence that by constant educational effort, both with the students in the medcial schools and with the practising profession through the avenue of the medical associations, a sympathetic understanding of the social and economic problems which necessitate social insurance can be brought about. The profession at large, he is certain, will learn that liberty can be preserved only by self-discipline and gradually will accept voluntarily what he deems their moral and social duty of lowering their fee schedules in treating insured families sufficiently to bring adequate medical care within the reach of all. This, the French health insurance scheme at present confessedly fails to do.

In addition, through a nationalization of some such collaboration arrangement as that which functions at Nancy, he anticipates that the cause of preventive medicine will be linked to the health insurance scheme.

As might be expected, the doctors who are performing the *contrôle médical* function of the funds are more critical of the French medical benefit arrangements. They are insistent, moreover, that the medical profession itself, determined as it is in defense of the medical benefit scheme which it has created, has not found in it the contentment which was anticipated. A statement made by the director of *Contrôle Médical* for a large regional union of funds, epitomizes this attitude: "Do not look for happy practitioners in France such as the panel doctors you find in Britain, for doctors cannot be at peace with the French arrangements."

The *contrôle* doctors seem in hearty agreement that, aside from their administrative problems, born of the indirect payment of medical bills through "reimbursement" to the patient, fundamental difficulties arise from payment by fees for each item of medical service. They believe that the Manchester experience is typical of fee system results, i.e. that prevention of imposition upon the funds by the irresponsible medical men is impracticable, however tight the control, and that the fee system under health insurance therefore favors the "chiseller" at the expense of the decent practitioner.

That the British open panel system, with remuneration on a capitation basis, supplemented, however, by a small *ticket modérateur*, would be a desirable arrangement not only for the funds but also for the medical man, was a conviction voiced frequently by *contrôle* doctors. Both lay and medical administrators in France agree with practitioners, however, that a small *ticket modérateur* would always be necessary in any scheme which provided family benefits. Apparently this is principally because of the belief that neurotic housewives otherwise would entertain themselves by going from doctor to doctor.

A basic distrust of the fee system of payment was also voiced by one of France's most distinguished psychiatrists, Dr. Jean Lépine.[1] Like Dr. Parisot, Dr. Lépine believes that the successful functioning of health insurance presents an essentially moral problem, which will be resolved only by continued educational effort. The insured must be taught, he pointed out, to comprehend the insurance principle if the social insurance system is to function honestly and effectively. They must learn that what their contributions are purchasing is *protection against risk*, and that the social insurance institution is not an automatic machine into which you put a coin and from which you must get a standard definite amount back.[2] The doctor for his part must be taught to appreciate the worker's need of health insurance protection which yields both cash and medical benefits. He must be prepared to do his share in a new organization of society to which such social services as health insurance are vital. His health insurance duties, an important one of which, emphasizes Dr. Lépine, is certification of his patient's need of cash benefit on disability, are a part of his medical job in this new society, and he must not shirk these duties.

Differing from his eminent colleague in Nancy, however,

[1] Dean of the Medical School of the University of Lyons.

[2] The authorities in this country who are struggling with the administration of recently established unemployment insurance schemes have a similar educational task on their hands.

Dr. Lépine believes that the French scheme of payment of the health insurance doctor (i.e. on a fee basis directly by the insured patient, with reimbursement from the fund) is fundamentally ill advised. To quote his words: "It must be remembered that men are frail and erring animals and it does not do to expect education in social and moral responsibilities to be able to make workable a system that invites its own breakdown. This organization of payment[3] out of insurance moneys for each medical act through the insured patient to the doctor, instead of directly to the doctor from the funds and on some basis less susceptible of abuse, facilitates the greatest amount of fraud and dishonest dealing and is an unintelligent organization."

Dr. Lépine believes that the experience of such experiments as that of the organization of maternity service *contrôle* by the Franco-American Foundation in Lyons, which, it will be remembered, pays the specialists at a fixed session rate, will point the way to change.

In leaving this question of contemporary medical appraisal in France of French medical benefit, it should be reiterated that the voices raised in advocacy of basic change are, at least in organized medical circles in France, overwhelmingly outnumbered by those which register definite antagonism to any such regulated situation as that of the health insurance doctor in Britain and in Denmark.

[3] i.e. the French health insurance scheme. This statement is quoted from an interview with Dr. Lépine in August 1936.

POSTLUDE

FOR completeness of medical benefit, Denmark, with nearly fifty years of health insurance behind her, stands out conspicuously in the three health insurance schemes reviewed. The Danish worker and his children, through guarantee of both medical and hospital care, are protected to an enviable degree.

Britain has evolved in her twenty-five years of health insurance a service, which is, within its prescribed general practitioner limits, of equal effectiveness. To this service, however, only the wage earner himself is eligible.

French medical benefit, in contrast, provides no actual services, medical or institutional. Instead, all costs of illness are put on a partially insured "reimbursement" basis, which in the evolution of French health insurance has remained practically unaltered. This limited protection is extended to the family as well as to the worker himself.

The doctor's position in the three health insurance schemes presents aspects of both similarity and contrast. The most important point of similarity is that all three schemes aim to conserve certain essential elements of "private practice." Thus they all accept the "private practice" precept that knowledge by the doctor that his income will suffer if he fails to satisfy his patients provides a sound disciplinary element which keeps the doctor on his toes. All three accept the private practice "free choice" theory that selection of the doctor gives the patient a trust and confidence in his physician which has definite therapeutic value.

Indeed, the organization of medical service in each instance stands upon the foundation principle that the traditional free doctor-patient relationship is a desirable one and that, as economic conditions frequently deny it to workers and their families, health insurance should restore it to them and to the doctors who would like to serve them. This fact needs emphasis, perhaps, because of the prevalent mistaken belief that health insurance, like many schemes of straight socialized medicine, aims at substituting for the private practice relationships a situation comparable to that of the teacher and student, i.e. a socialized medical service rendered by a closed technical staff which is selected by administrative authorities and whose remuneration has no direct relation to the ability to attract and hold patients.

Another point of similarity in the three schemes is that in each country collective action by the organized medical profession plays a vital part in determination of the health insurance doctor's privileges, responsibilities, and remuneration.

Thus in Britain the Insurance Acts Committee of the British Medical Association (accepted as agent for the health insurance doctors) consults with the Ministry of Health regarding the capitation fee and mileage rates to be paid, discusses with the Ministry the promulgation of all medical regulations, and collaborates with, and has representation in, the Ministry in the functioning of the disciplinary machinery affecting the doctors.

In Denmark the organized profession plays an even more significant rôle. Collective bargaining agreements with the health insurance organizations directly determine, not merely influence, all conditions of health insurance practice, except that of the ultimate appellate "disputes" machinery which is set by law. These agreements require membership in the medical association for entering health insurance practice; settle all conditions of health insurance medical work, including privileges, duties, and both the amount and method of remuneration; arrange for conciliation and primary arbitration procedure in disputes which concern the medical profession.

In France, while collective agreements, as in Denmark, determine most of the duties and privileges of the doctor, including the reimbursement *tarifs* which affect medical income, these agreements are not completely free as in Denmark, for their content is, to all practical purpose, largely prescribed by the governmental authorities. The organized profession, however, participates in the formulation of governmental regulations and is represented in all governmental agencies which administer the social insurance laws. Moreover, almost all the basic disciplinary procedure is exclusively in the hands of the organized profession.

In all three schemes the doctor's obligations, whether legally imposed or contractually self-imposed, include not only medical care and prescription for the insured but also certification of his capacity or incapacity for labor, to guide the administration of cash benefits. This certification duty is accepted as the inevitable responsibility of the attending physician, a "prescribing of rest" so to speak, which is essential to the treatment of disabling illness.

The chief points of dissimilarity are related to the insurance remuneration of the practitioner. In Britain he is paid from a governmental pool; in Denmark from the exchequer of a health insurance society or group of societies; in France the doctor primarily collects his bills in toto from his patients, but ultimately, through the patient's medical benefit right to partial "reimbursement," partly from the insurance funds. In both Britain and Denmark, no choice either as to method or amount of remuneration is permitted the *individual* practitioner, while in France each doctor may bill his patient as he chooses and collect what he can.

In Britain the practitioner is paid exclusively on a capitation basis. In Denmark his remuneration is on a capitation basis in some regions, and in others on a fee basis, with a definite trend toward the capitation basis. In France the doctor is paid theoretically in all instances on a fee basis. Practically, through the growth of dispensaries and clinics, to a certain and increasing extent he is paid on a salary or "session" basis. Almost never is any of his remuneration on a capitation basis.

In respect to the right to enter health insurance practice and to the supervisory contact of the health insurance organizations with the doctors, there is interesting contrast in the arrangements of the three countries. The British law opens health insurance practice to the entire profession, makes the doctor answerable to the health insurance authorities for a decent performance of his health insurance work, and frees him from all contact with the insurance societies. A supervisory medical referee service exists, which is purely governmental and wholly disassociated from the societies.

Under the Danish law, in contrast, the health insurance societies are authorized to arrange medical service as they choose. Actually, however, individual medical contracts are not made, most health insurance practice is "open," and its terms are set in collective contracts between the medical associations and regional federations of societies. While the societies legally could set up the most complete medical supervisory machinery, none whatever has been created.

In the French scheme, on the other hand, all doctors may treat insured persons. The doctor is legally free of any required contact either with the insurance funds or with the government. Yet, and indeed *because* of this situation, there is more critical review of his work than in Britain, with its official referee service. In France, moreover, this review is made by medical employees of the funds themselves, who, through the diagnostic work of *contrôle médical*, check up, on a constantly increasing scale, the medical practitioner's health insurance work.

It is the belief of the medical profession in both Britain and Denmark that health insurance gives them a degree of professional freedom[1] in the practice of medicine among the mass

[1] This freedom is, of course, much more complete in Denmark than in Britain, because of the far more complete medical benefit of the Danish scheme. cf. p. 159 *supra*.

of the people such as is enjoyed in private practice only in treating the well-to-do. The members of the profession find this good. In addition, experience has proved to them that they make both a better and a steadier income under health insurance than they did without it. They find this also good.

They apparently feel that the regulated aspects of their health insurance practice, which is the price they pay for its advantages, is an inevitable and not very onerous *quid pro quo*. They disagree with the assumption of their French professional brethren that substantial advantages for the insured population and for the medical profession can be obtained from health insurance without changing the private practice methods of collecting medical bills.

The French doctors, on the other hand, although perturbed at some of the developments which their organization of medical benefits has stimulated, still are, in general, advocates of their "free position." Whether or not the medical ventures based upon opposition to this position are merely the exceptions that prove the rule or are prophetic in their significance is, at present, merely a matter of conjecture.

Most physicians in the health insurance countries surveyed, stress the fact that health insurance has greatly improved the practitioner's chance of preventing the invalidity of his patient. Because of the severity of the European tuberculosis problem, medical men abroad are especially impressed with the contribution which health insurance has made to the handling of tubercular wage earners. Their opinion is well summed up in the following words[2] of the distinguished exponent of social medicine, Dr. Réné Sand:

"Only through health insurance is the medical profession furnished with the 'sinews of war' necessary for the successful combating of tuberculosis. These sinews of war are vital

[2] Quoted from an interview with the author in August 1936 in Brussels where Dr. Sand was serving as adviser to the Belgian Ministry of Health. Dr. Sand (known for his long connection with the International Labor Office and for his writings in the field of social medicine) is one of the outstanding European tuberculosis specialists.

not only for purchasing the medical and institutional care of the 'cures' but also to provide cash benefits for the family's support, without which the doctor cannot hope to induce the worker to take the rest and care which spells recovery in the incipient tubercular case."

In conclusion, it merits emphasis that health insurance, by throwing a constant spotlight on the extent of illness, its neglect, and its enormous cost, is promoting the cause of preventive medicine in all fields. That this cause must become the concern of the individual practitioner, if the traditional doctor-patient relationship is to survive, is a growing conviction of the leaders in organized medical circles in France, as in Britain and Denmark. This conviction is well expressed in an editorial in the *Rhone Medical Journal* penned by the 1936 president of the French Medical Association[3] under the title "At the Crossroads":

"Events are taking place at a rate that could not have been foreseen. Those that flattered themselves that they would direct them find themselves obliged to follow them instead.

"At this point it behooves all of us who would preserve the free practice of medicine to take a clear view of the situation.

"The time is over for academic discussions.

"Will we cling tenaciously and hopelessly to the classic conception of the practice of our profession, or will we have the wisdom to realize that hitherto we have visualized a far too limited function for our profession? If we select the first alternative, then, I will state bluntly, it is the end of our profession. So much the pity for those who close their eyes to this fact. . . . They will have been the principal builders of their unhappy fate.

"If we select the second alternative, if we realize that our function is not confined to the individual treatment of a sick person, if we can see that we must concern ourselves with social planning, there will yet be happy days for the profes-

[3] Dr. Joseph Chalier of Lyons. See *Bulletin Officiel du Syndicat des Médecins du Rhône*, July 1936, pp. 390-1.

sion of medicine, and physicians will enjoy the gratitude and esteem of those who devote themselves to general hygienics and to the future condition of the French people. We can and must become both the counsellors and technical agents of the public authorities and the great organizations which devote themselves to improving the public health. Let us learn to collaborate. Let us learn to rid ourselves of our far too narrow conception of the proper rôle of the medical profession. 'Healers of the sick' is, of course, a glorious title, which still should command respect.

"But let us not keep in sight merely the maintenance of our traditional Charter of Medicine.

"The Charter's terms will be the better accepted the more we set ourselves everywhere throughout the country to developing and evolving and to aiding with our technical skill the cause of general prevention and early diagnosis of disease."

INDEX

INDEX

ADDENDA

p. 146, *after* l. 2, *add:* Mileage is paid in the amout of 7/10 krone per kilometer for each kilometer of rural travel in excess of two, and for each kilometer "off the regular roads."

p. 170, *after* l. 22, *add:* The insured person now is entitled on unemployment to have credited to his insurance account a contribution of thirty francs a quarter. This is granted for not more than two consecutive quarters in the year.